Applied Attention Theory

Christopher D. Wickens and Jason S. McCarley

CRC Press
Taylor & Francis Group
Boca Raton London New York

CRC Press is an imprint of the
Taylor & Francis Group, an **informa** business

CRC Press
Taylor & Francis Group
6000 Broken Sound Parkway NW, Suite 300
Boca Raton, FL 33487-2742

International Standard Book Number-13: 978-0-8058-5983-6 (Softcover)

Library of Congress Cataloging-in-Publication Data

Wickens, Christopher D.
   Applied attention theory / authors, Christopher D. Wickens, Jason S. McCarley.
      p. ; cm.
   Includes bibliographical references and index.
   ISBN 978-0-8058-5983-6 (hardcover : alk. paper)
   1. Attention--Physiological aspects. 2. Cognitive neuroscience. I. McCarley, Jason S. II. Title.
   [DNLM: 1. Attention--physiology. 2. Cognitive Science--methods. 3. Psychological Theory. BF 321 W636a 2007]

QP405.W55 2007
612.8'2331532--dc22   2007030245

Visit the Taylor & Francis Web site at
http://www.taylorandfrancis.com

and the CRC Press Web site at
http://www.crcpress.com

Contents

Preface

We wrote this book in an effort to relink two areas of study that arose at the same time and are of the same ancestry—the pioneering human performance research of Cherry, Fitts, Broadbent, and Mackworth—but that soon grew aloof from one another. One was the basic study of attention, in which we saw sophisticated and exciting theories of how the senses gather and the brain processes multiple streams of information but too little concern for search, display processing, and multitasking in the complex world beyond the laboratory. The other was the study of human behavior outside the laboratory, in which we found studies of complex displays, inattentive drivers, and overloaded pilots but too little consideration of the elegant theoretical work that might provide a basis for predicting the successes and failures of our real-world attentional endeavors. With this book, we have tried to reunite these two areas, identifying correspondences and complements, and we hope to stimulate other researchers to do the same.

In the process of completing the book, we have accumulated a debt of gratitude for Jane Lawhead, whose organization, coordination, and hard work helped us to assemble the various components and chapters. We also extend our great thanks to the University of Illinois and Alionscience: MA&D operations for providing the resources and access to literature as we wrote and to Art Kramer and Wai-Tat Fu for their helpful comments on chapter drafts. Finally, we gratefully acknowledge the support offered by our wives, Linda and Cyndie, and our dogs, Mattie and Buster, throughout the project.

Authors

Chris Wickens received his bachelor's degree in physical sciences from Harvard College in 1967 and his Ph.D. in experimental psychology from the University of Michigan in 1974. From 1969 to 1972 he served in the U.S. Navy. He has been a professor of psychology at the University of Illinois since 1973 and a professor of aviation since 1983. From 1983 to 2005 he served as head of the Human Factors Division at the University of Illinois. He has authored two textbooks in engineering psychology and in human factors and is the author of more than 150 research articles in refereed journals. His research interests focus on applications of attention models and theories to the design of complex displays (particularly in aviation) and to the understanding of human multitasking capabilities. He is a fellow of the Human Factors & Ergonomics Society and has received the society's annual award for education and training of human factors specialists, along with the University Aviation Association President's Award, (2005), Flight Safety Foundation Airbus Human Factors Award (2005), Federal Aviation Administration Excellence in Aviation Award (2001), Henry L. Taylor Founder's Award, and Aviation, Space, & Environmental Medicine Human Factors Association Award (2000).

Jason McCarley received his bachelor's degree from Purdue University and his Ph.D. from the University of Louisville and then served as a postdoctoral scientist at the Naval Postgraduate School in Monterey, California, and the Beckman Institute at the University of Illinois. He is currently assistant professor in the Institute of Aviation and the Department of Psychology at the University of Illinois. His research interests focus on basic and applied issues in visual perception, attention, and cognition.

1

Attention

Introduction

The study of attention has importance that is at once historical, theoretical, and applied. From a historical perspective, over a century ago William James, the founder of American psychology, devoted a full chapter to the topic in his classic textbook, *Principles of Psychology* (1890). Interest in attention as a field of psychological study waned during the behaviorist period in the first half of the century when attention was improperly dismissed as a mediating mental variable that could not be directly measured and was therefore outside the bounds of scientific inquiry. Nevertheless, even during this time a few classic studies such as Jersild's (1927) and Craik's (1947) work on attention switching were published. Then, following World War II interest in the topic blossomed, as is discussed in the next chapter, so it remains a fundamental element of psychological research to this day.

The theoretical importance of attention can be seen at two different levels. First, as one of the three main limits on human information processing (along with storage–memory and speed–response time), the study of attentional processing capacity is of fundamental interest in its own right: How many tasks can we do at once? How rapidly can we switch from task to task? How widely can we deploy attention across the visual field? Second, attentional properties underlie many other psychological phenomena: Attention is necessary to hold information in working memory and to efficiently move information to long-term memory, that is, to learn it. It is a vital component underlying decision making and is integrally related to perceptual processing.

The applied importance of attention is also manifest in several ways. For example, the aforementioned fields of memory, learning, and decision making all scale up to real-world problems such as eyewitness testimony, training, choice, and display design, so the attentional components underlying these naturally scale up as well. But attentional challenges and issues also are directly relevant outside of the laboratory: The dangers of distracted drivers, the attentional overload of making sense of massive data bases, the rapid attention switching required in the electronic workplace, the success or failure of unreliable alarms to capture attention, and the behavior of children with attention deficit disorder are all examples.

Varieties of Attention

The word *attention* encompasses a broad array of phenomena. Another founder of American psychology, Titchner (1908, p. 265), in introducing his chapter on attention noted, "The word 'attention' has been employed in the history of psychology to denote very different things" and then went on for a paragraph to list these different meanings. The present book considers several varieties of attention (see also Parasuraman and Davies 1984), which can be illustrated within the context of highway driving. The driver will first want to concentrate, or focus attention, on the driving task, in the face of many possible distractions—competing tasks and nonrelevant events. Thus, focused attention can apply to a task or a particular channel or source of environmental information. Rarely, however, does the driver carry out only one task, but he or she will often select between alternatives—for example, between lane keeping and checking the map. Here, selective attention can be defined either at a gross level, as selecting to devote attention to one task or another, or at a fine level, usually represented by visual scanning, as looking from one place to another. Intrinsic to the concept of selective attention is the notion of an attention switch, which describes the process of moving attention from one task, or channel, to another.

Now, sometimes the driver can succeed at actual multitasking, not by switching between tasks or channels but by actually processing them in parallel (i.e., simultaneously). Here we speak of the success of divided attention, which, like selection, can be described at two levels. At the task level, successful divided attention may include such things as successful lane keeping while understanding a news program or appreciating music on the radio. At the perceptual level, divided attention may involve the parallel processing of two aspects of a stimulus; a quick glance at the well-designed map, for example, can reveal the length of the highway, its direction, and destination, all understood within the single perceptual experience. Correspondingly, a quick glance out the windshield can reveal the color of an oncoming car, its speed, and its heading, all at the same time. It should be noted, of course, that parallel processing or successful divided attention does not necessarily imply that either of these are perfect—only that they are better (usually faster) than a purely serial allocation of selective attention would allow.

Finally, it is possible to speak of sustained attention as that variety of attention mobilized in continuous mental activity, whether that activity appears to be of high complexity (e.g., completing a three-hour final exam) or low (e.g., maintaining the night watch); in either case, there is a toll on human cognition due to trying to mobilize the high effort to perform a task for a long period of time.

In addition to these five varieties of attention—focused, selective, switched, divided, and sustained—it is possible to conceptualize attention metaphorically in two different ways: as a mental filter and mental fuel, both shown in Figure 1.1. The figure represents a conventional information processing

A Simple Model of Attention: The Filter and the Fuel

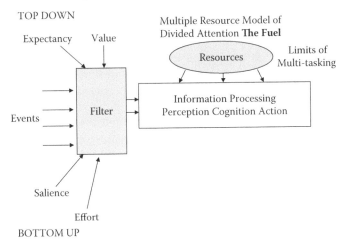

Figure 1.1 A simple model of attention. The influences on the filter of selective attention are shown, as are the influences of limited resources on the information-processing activities involved in divided attention and multitasking.

model of human performance and highlights the more perceptual aspects of attention as being the front-end *filter*, selecting certain stimuli or events to be processed and filtering others out as less relevant. If selection is aided by divided attention, then two or more channels may gain access to the filter at once. Then, as information-processing activities are carried out on the selected environmental information or on self-generated cognition, limits of the mental resources, or fuel, that supports such processing constrain the number of processes that can be carried out at once. This is true whether in the service of a single task (e.g., holding subsums while doing mental multiplication) or in the service of multitasking. Figure 1.1 presents a view of attention that is simplified but provides a heuristic foundation on which we build throughout the chapters of the book

Relation to Other Applied Domains

As has been noted already, attention is closely tied to other domains of cognitive psychology, such as memory, learning, and decision making and perception. However, attention also links closely with four important applied domains, which are described briefly here and are revisited at times later in the book.

First, in the study of human error (Holnagel 2007; Reason 1990; Sharit 2006) attentional errors, particularly attentional lapses, represent a major important source of cognitive deficit. Second, the lapses of attention are closely related to aspects of situation awareness (Durso et al. 2007; Endsley 1995, 2006;

Tenney and Pew 2007), a concept that has gained and has warranted great popularity within the past two decades. Situation awareness can be characterized as "an internalized mental model of the current state of the operator's [dynamic] environment" (Endsley 2006, p. 528). In particular, it is conventional to break situation awareness into three levels or stages: (1) noticing events in the dynamic environment; (2) understanding the meaning of those events; and (3) predicting or projecting their implications for the future. Here, stage 1, noticing, clearly depends on effective deployment of the attention filter. Stages 2 and particularly 3 are resource intensive, if it is done well: An operator who fails to anticipate future events because of high concurrent task load will not succeed well in dealing with those future events.

Third, the study of mental workload has been another popular area of applied interest within the past twenty years, as designers endeavor to create systems in which the human mental resources required to operate these systems remain less than the available resources; that is, the fuel in Figure 1.1 is not depleted by mental operations so that there is always some residual in the tank to handle unexpected emergencies. Thus, the study of mental workload calls for ways to measure the fuel—both expended and remaining—and to reduce the demands imposed on it.

Fourth, with the increasing sophistication of computers and technology, often created with the goal of reducing mental workload, has come the issue of human–automation interaction (Sheridan 2002; Sheridan and Parasuraman 2006). Two aspects of attention in human–automation interaction are directly linked to situation awareness and workload as described already. First, a major impetus for introducing automation in the first place is the desire to reduce workload. At the individual human level, this may be to prevent human operator workload overload, as when an autopilot can offload the pilot allowing her to concentrate on many other task responsibilities. At a level of macroergonomics, automation and workload are related by manpower concerns: If the function of one member of a three-person crew is automated, an operation can then be conducted that has a 33% reduction in personnel costs. Second, it is now well established that high levels of automation that are designed to reduce workload also reduce operator awareness of the processes that are automated but over which the human may still have full responsibility. Such a reduction in situation awareness is often mediated directly by attention, as the human may cease paying much attention to the automated processes—a potentially catastrophic behavior if the automation fails or if other processes under control of the automation go wrong.

Scaling Up Basic to Applied Research

As noted already, attention is a psychological construct that binds basic and applied issues. In this regard, the reader will note the effort made here to treat both endpoints of this continuum as well as the full range of studies and issues in between. If both endpoints of this continuum are examined, at one end is seen elegant and well-controlled theoretical work, conducted with no specific application in mind and, because of high control, often showing effect sizes that are no larger than a few milliseconds but highly significant in statistical terms. At the other end are seen analyses of many real-world accidents or incidents that are clearly related, in part, to attentional breakdowns (e.g., 10–50% of automobile accidents are related to distraction [Weise and Lee 2007]). But because of the lack of control over the collection of such data, it is impossible to preclude other contributing causes or to draw strong causal inferences.

The challenge in engineering psychology, the domain that represents the spirit of this book is to link the two endpoints. How do we extract the theory-based attentional phenomena and identify which ones scale up to account for real-world attentional failures and successes? How much variability do these well-controlled phenomena account for in the world outside the laboratory? Most critically, how well can the independent variables that modulate the strength of attentional effects in the laboratory be captured by design and training variables in the real world of human interaction with complex dynamic systems, where attention plays such a vital role?

An issue that grows in importance as one moves from more basic to more applied research is the distinction alluded to previously between statistical significance, typically defined in terms of the p-level of a statistical test, and practical significance, defined by the size of an effect in raw units (Wilkinson 1999)—for example, one second saved in the braking time in an automobile— by adopting a recommendation from attention research, like providing a head-up display or auditory display, or a 20% reduction in accident rate. Whereas basic research is typically most concerned with statistical significance, applied research must give equal concern to both types. After all, an intervention that only saves 1/100 of a second in braking response time in the automobile will be of little benefit, even if it may be significant at a $p < .05$ level. On the other hand, a manipulation that offers a potential one-second savings may be of tremendous importance to more applied researchers even if it has not quite reached the conventional $p < .05$ level effect of statistical reliability (Wickens 1998) but still is close to that level. In this regard, it is important to note that applied researchers cannot afford to ignore statistical significance—only that they must temper their concern for statistical significance with an appreciation of practical significance.

Importantly, there are two phases to the process of the scaling-up findings from the lab to application. First, it is important to show how laboratory phenomena in attention express themselves in real-world scenarios. Second, but equally important, is to devise attention-based solutions to enhance productivity and safety in these environments. The following chapters try to show how this is done by integrating across the basic-applied continuum.

The Role of Models

An aspect of the efforts to transition from basic research to applications that has gained increasing importance is the development of human performance and cognitive models (Foyle and Hooey 2007; Gray 2007; Laughery, LeBiere, and Archer 2007; Pew and Mavor 1998). Such models can be of two general classes. Descriptive models, like the bottleneck model shown in the next chapter, describe the mediating processing mechanisms of attention in a way that accounts for performance qualitatively. Sometimes descriptive models can spawn computational models, which can produce quantitative predictions of actual performance measures such as attention-switching time or probability. These are of great value because they may be able to predict safety measures like the one-second brake time savings discussed previously. Also, paralleling the distinction made with regard to attention research is one of basic versus applied computational models. Basic models tend to be more exacting, requiring a fairly high bar for validation. Applied models, on the other hand, tend to be somewhat less accurate but more encompassing of a wider range of conditions and environmental variables.

The great value of applied computational models, once they are validated by human performance data, is that they often allow designers to know that a system is likely to be unsatisfactory to human operators even before the system has been put into production or even mocked up for human testing. For example, a model might be one that predicts that drivers will need to bring their eyes off the roadway for at least two consecutive seconds to operate a particular piece of in-vehicle technology. Such a long exposure to roadway hazards might be viewed as an unacceptable compromise of highway safety and might force a redesign of the interface before it is introduced into the vehicle.

The chapters of the book, in which these varieties of attention are covered from both a theoretical and applied perspective, are organized as follows. Chapter 2, "Single-Channel Theory and Automaticity Theory," describes two of the most important historical concepts in attention and as a consequence provides some historical context as well as defines two elements that underlie many of the concepts that follow.

Chapter 3, "Attention Control," chapter 4, "Visual Attention Control, Scanning, and Information Sampling," chapter 5, "Visual Search," and chapter 6, "Spatial Attention and Displays," focus on different aspects of visual attention, as this research has dominated much of attention theory over the past three to four decades and has proven of vital importance in many high-risk activities beyond the laboratory, where the visual world of displays and events is often of overwhelming complexity.

Chapter 7, "Resources and Effort, chapter 8, "Time Sharing and Multiple Resource Theory," and chapter 9, "Executive Control Attention Switching, Interruptions, and Workload Management," address different issues of multiple-task performance, move beyond the visual perceptual world, and create the foundations of a general model of how tasks interfere or compete with each other when people must divide attention between them: interference caused by their difficulty and their similarity. In addition, chapter 5 also highlights the critical role of effort in single task behavior, such as decision and choice, to the extent that humans tend to be effort-conserving in their choice of activities. Chapter 9 examines how multiple tasks are managed in a discrete fashion.

Chapter 10, "Individual Differences in Attention," considers the role of ability differences, age differences, and training differences in attention with emphasis on the real-world implications of such differences to areas such as training, licensing, selection, and user support. Chapter 11, "Cognitive Neurosciences and Neuroergonomics," focuses on brain mechanisms underlying attention but does so in a way that the practical applications of these mechanisms are clear.

Within each chapter, we try to maintain a balance between theory and applications, although occasionally we may veer more one way than another. Our hope is that the book will stimulate basic researchers to understand the importance of their work in the world beyond the laboratory and perhaps to goad them into tackling some of the complex problems that exist there—driving and cell phone usage has been a beautiful success story in this endeavor. Equally, we hope that those involved in applied human factors area of design and measurement can appreciate the importance of attention and the value of good science underlying attentional phenomena in exercising their careers.

2

Single-Channel Theory and Automaticity Theory

Introduction

The concept of attention is often invoked to explain an inability to do more than one thing at a time. Such a statement represents an intuitive description of the more formal psychological construct of single-channel theory (Broadbent 1958; Craik 1947; Welford 1952, 1967, 1968), or single-channel behavior, an extremely constraining view of the capabilities of attention. Very simply, such a theory predicts that the time required to perform any two tasks together will be at least as great as the sum of the time required to do each in isolation.

At the other end of the capabilities spectrum is the idea that attention can sometimes allow performance of multiple tasks or process multiple channels of information simultaneously and without interference—that is, in parallel and with unlimited capacity—as if performance were automated, requiring little or no attention at all. Though it is obvious that none of us can parallel process all possible task combinations, certain circumstances do support effective parallel, or automated, processing, just as others require nearly complete single-channel processing. The purpose of this chapter is to introduce the historical and theoretical foundations of each of these two diametrically opposed characterizations of attention support for information processing, as their origins date to well before attention research became popular. Indeed, over a century ago, James (1890, p. 409) invoked both concepts within the sentence: "If you ask how many ideas or things we can attend to at once, the answer is not very easily more than one, [single-channel theory], unless the processes are very habitual [automaticity theory]." In describing each of these concepts, we then show how the extreme versions can be tempered to describe real-world behavior that is neither completely serial (single channel) nor completely parallel, and we invoke the construct of mental effort to span the range between these two endpoints. Although both a fully serial and an unlimited capacity parallel model can be rejected as a general descriptions of human behavior and cognition, both are valuable (1) as anchors of a continuum against which other types of behaviors can be contrasted, and (2) as accurate characterizations of some forms of human behavior that systems designers should seek to avoid (i.e., single-channel behavior) or attain (i.e., automaticity). Our approach focuses on three historic paradigms of attention research: (1) the psychological refractory period (PRP) paradigm, which served as the foundation for single-channel theory; (2) the auditory shadowing paradigm, which helped reveal where the single-channel bottleneck might be located (and, hence, softened the single-channel theory's extreme rejection of human multitasking capabilities); and (3) the visual search and time-sharing paradigms that produced evidence for the possibility of automaticity in cognition and behavior.

Single-Channel Theory and the Psychological Refractory Period

Many of the historical roots of single-channel theory are planted within the paradigm of the psychological refractory period, or PRP (Kantowitz 1974; Meyer and Kieras 1997a, 1997b; Pashler 1998; Telford 1931). In a PRP task, the subject is asked to make separate, speeded responses to a pair of stimulus events that are presented close together in time. The separation in time between the two stimuli is called the *stimulus onset asynchrony* (SOA). The general finding is that the response to the second stimulus is delayed by the processing of the first when the SOA is short. Suppose, for example, a subject is presented two stimuli: a tone (S_1) and then a light (S_2), separated by a variable SOA. The subject has two tasks. The first (Task$_1$) is to respond by pressing a key (R_1) as soon as the tone is heard. The second (Task$_2$) is to respond by speaking a word (R_2) as soon as the light (S_2) is seen. If the SOA between the two stimuli is sufficiently short (i.e., less than the response time to the first stimulus), then response time to the light (RT_2) will be prolonged. RT to the tone (RT_1), in contrast, will generally be unaffected by the second task. The PRP delay in RT_2 is typically measured with respect to a single-task control condition in which the subject responds to S_2 while simply ignoring S_1.

The most plausible account of the PRP proposes the human being to be a single-channel processor of information. The single-channel theory of the PRP was originally proposed by Craik (1947) and has subsequently been expressed and elaborated on by Bertelson (1966), Davis (1965), Kantowitz (1974), Meyer and Kieras (1997a, 1997b), Pashler (1989, 1998), and Welford (1952, 1967). Such a view is compatible with Broadbent's (1958) conception of attention as an information-processing bottleneck that can only process one stimulus or piece of information at a time. In explaining the PRP effect, single-channel theory assumes that the processing of S_1 temporarily occupies the single-channel bottleneck of stimulus processing. Thus, until R_1 has been released (i.e., the single channel has finished processing S_1) the processor cannot begin to deal with S_2. The second stimulus, S_2, must therefore wait at the gates of the bottleneck until they open. This waiting time is what prolongs RT_2. The sooner S_2 arrives, the longer it must wait. According to this view, anything that prolongs the processing of S_1 will increase the PRP delay of RT_2. Reynolds (1966), for example, found that the PRP delay in RT_2 was

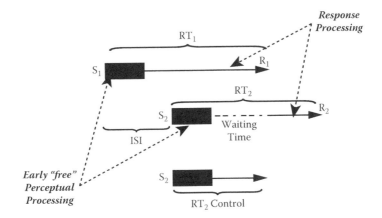

Figure 2.1   Single-channel theory explanation of the psychological refractory period. The figure shows the delay (waiting time; the dashed line) imposed on RT_2 by the processing involved in RT_1. This waiting time makes RT_2 in the dual-task setting (top) longer than in the single-task control (bottom).

lengthened if the RT task of RT involved a more complex choice rather than a simple response. This simple relationship between interstimulus interval (ISI) (and its inverse, waiting time) and RT_2 is represented graphically in Figure 2.1, importantly indicating the one-for-one exchange between the two variables—that is, the earlier you come, the longer you wait.

Within the PRP paradigm, the bottleneck in the sequence of information-processing activities does not generally appear to be located at the peripheral sensory end of the processing sequence (like blinders over the eyes that are not removed until R_1 has occurred). If this were the case, then no processing of S_2 whatsoever could begin until RT_1 is fully complete. However, much of early sensory and perceptual processing is relatively automatic (Posner, Snyder, and Davidson 1980). The processing bottleneck evident in the PRP, rather, appears to arise at the stage of response selection, the process of mapping a stimulus to the appropriate behavior. Therefore, the basic perceptual analysis of S_2 can often proceed even as the processor is fully occupied with selecting the response to S_1 (Karlin and Kestinbaum 1968; Keele 1973; Pashler 1989, 1998). Once perceptual processing of S_2 is completed, however, response selection for T_2 apparently has to wait for the bottleneck to dispense with R_1. These relations are shown in Figure 2.2.

Imagine as an analogy two patients visiting the same physician for a physical. The total time at the doctor's office for the first patient is the sum of the time it takes to fill out paperwork and be seen by the doctor. This time remains the same even if a second patient arrives soon after the first. The total time of the second patient's office visit, however, is determined also by how soon he arrives after the first patient. If the second patient arrives immediately after the first, then the two can fill out the paperwork concurrently.

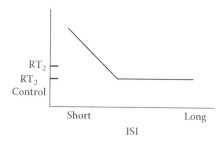

Figure 2.2 Hypothetical graph showing data predicted by single-channel theory, plotting RT_2 (response to the second arriving stimulus) as a function of ISI, the delay in S_2 arriving after S_1. If this is short, S_2 must wait to be processed until R_1 is released. Every unit of time ISI is lengthened, and the wait time is shortened by the same amount—thus, the slope of –1.0 in the function. Eventually at an ISI about equal to RT_1, ISI is long enough so that S_2 arrives well after completion of R_1, and there is no delay in RT_2, relative to the control condition.

However, the second patient cannot see the doctor until the first patient's exam has been completed. After completing his paperwork he will therefore dawdle in the waiting room until the first patient has been discharged, and the length of his visit will be prolonged by exactly the length of the first patient's exam. Some of this wait time, however, for patient 2 can be absorbed by the time needed to complete paperwork, a task that can proceed in parallel with the doctor's examination of the first patient. Indeed, if the second patient's paperwork is completed at just the moment the first patient is discharged, there will be no extra waiting time penalty at all for patient 2. In this example, the doctor, who can examine only one patient at a time, is the single-channel bottleneck; the first- and second-arriving patients reflect S_1 and S_2, respectively, and the total visit times for each patient reflects RT_1 and RT_2.

Returning to the PRP paradigm, we see that the delay in RT_2 beyond its single-task baseline will increase linearly—on a one-to-one exchange basis—both with a decrease in ISI and with an increase in the time needed for T_1 response selection, since both of these manipulations increase the waiting time. Again, this relationship is shown in Figure 2.2. Assuming that the single-channel bottleneck is perfect (i.e., postperceptual processing of S_2 will not start at all until R_1 is released), the relationship between ISI and RT_2 will look like that shown in Figure 2.2. When ISI is long—much greater than RT_1—RT_2 is not delayed at all. When ISI is shortened to about the length of RT_1 some temporal overlap will occur, and RT_2 will be prolonged because of a waiting period. This waiting time will then increase linearly as ISI is shortened further.

The relationship between ISI and RT_2 as shown in Figure 2.2 describes rather successfully a large amount of the PRP data (Bertelson 1966; Kantowitz 1974; Meyer and Kieras 1997a, 1997b; Pashler 1989, 1998; Smith 1967). There are, however, three important qualifications to the general single-channel model as it has been presented so far.

(1) When the ISI is very short (less than about 1/10 sec), a qualitatively different processing sequence occurs; both responses are emitted together (grouping), and both are delayed (Kantowitz 1974). It is as if the two stimuli are occurring so close together in time that S_2 gets through the single-channel gate while the gate is still accepting S_1 (Kantowitz 1974; Welford 1952). Response grouping also obtains when the subject is required to execute two different responses (e.g., a key press and a spoken word) to a single stimulus (Fagot and Pashler 1992).

(2) Sometimes RT_2 suffers a PRP delay even when the ISI is greater than RT_1. That is, S_2 is presented after R_1 has been completed. This occurs when the subject is monitoring the feedback from the response of RT_1 as it is executed (Welford 1967) and may also characterize an attention-switching time penalty between tasks.

(3) The one-for-one exchange (–1 slope) between ISI and RT_2 is not always found (Kahneman 1974), with a shallower slope suggesting some parallel processing between the two (Dixon, Wickens, and Chang 2005; Wickens, Dixon, and Ambinder 2006).

Applications of Single-Channel Theory to Workload Prediction

Not surprisingly, single-channel theory has sometimes been called bottleneck theory, under the assumption that the single channel represents a bottleneck in information processing. Using various experimental paradigms, but particularly those of the PRP, theory-oriented investigators have provided conflicting evidence on whether the bottleneck of information processing is early (i.e., at perception), occurs at the later stage of response selection (Pashler 1991, 1998), or can perhaps under the appropriate circumstances be circumvented altogether (Schumacher et al. 2001). Such discussions parallel debates in the basic attention research around auditory shadowing, discussed following. The approach taken in the current chapter assumes that applications are best realized by considering the task and environmental circumstances that are most likely to produce single-channel behavior. In the real world, three diverse circumstances that are likely to produce single channel behavior are as follows.

(1) When visually displayed sources of information requiring foveal vision are widely separated (Dixon, Wickens, and Chang 2005; Liao and Moray 1993). Here the single channel is obviously preperceptual.

(2) When multiple tasks demand processing in rapid sequence—real-world analogies to the PRP paradigm, such as when a driver must

respond to a suddenly appearing roadway hazard, at the very instant that a passenger asks a question (see Levy, Pashler, and Boer 2006).
(3) When two tasks demand a relatively high level of cognitive involvement, as when trying to listen to two engaging conversations at once or when thinking, or problem solving, while reading an unrelated message.

Because these circumstances do represent meaningful samples of human behavior in many real-world contexts, it turns out that the single-channel model has had a very useful application in two important areas. One of these has been to identify the circumstances under which operators in safety critical tasks (e.g., driving, flying) tend to regress into some form of single-channel behavior akin to attentional tunneling or cognitive lockout (Wickens 2005a). For example, fault management appears to induce such behavior (Moray and Rotenberg 1989), as does the extreme interest and engagement fostered by certain tasks like cell phone conversations while driving (Horrey and Wickens 2006; Strayer and Drews 2007; Strayer, Drews, and Johnston 2003). These issues are discussed further in chapter 9.

The second important application area of single-channel processing is in workload modeling and workload prediction (Hendy, Laio, and Milgram 1997; Laughery, LeBiere, and Archer 2006; McMillan et al. 1989; Sarno and Wickens 1995). Such models are designed to make both absolute and relative predictions of how successfully people will perform in multitask settings.

The goal of absolute predictions of task interference and workload calls to mind the kind of question asked by the Federal Aviation Administration before certifying new aircraft: Are the demands imposed on the pilot excessive? (Ruggerio and Fadden 1987). If *excessive* is to be defined relative to some absolute standard, such as 80 percent of maximal capacity, an absolute workload question is being asked. Practitioners speak of a red line above which workload should not be allowed to cross. A common approach to absolute workload and performance prediction is timeline analysis, by which the system designer constructs a temporal profile for the workload that operators encounter during a typical mission, such as landing an aircraft or starting up a power-generating plant (Kirwan and Ainsworth 1992). In a simplified but readily usable version compatible with single-channel theory, it assumes that workload is proportional to the ratio of the time required (TR) performing tasks to total time available (TA). If one is busy with some measurable task or tasks for 100 percent of a time interval, workload is 100 percent during that interval. Thus, as shown in Figure 2.3, the workload of a mission would be computed by drawing lines representing different activities of length proportional to their duration. The total length of the lines within an interval would be summed and then divided by the interval time (e.g., 10 seconds) (Parks and Boucek 1989). In this way, the workload encountered by or predicted for different members of a team (e.g., pilot, copilot, and flight engineer) may be compared and if necessary tasks can be reallocated or automated. Furthermore, epochs of peak workload or work overload, in which

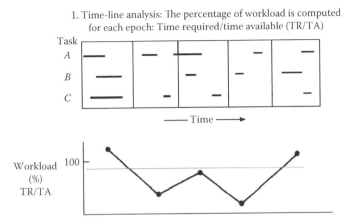

Figure 2.3   Timeline analysis of workload based on single-channel theory. The task timeline is shown at the top of the figure. Assuming that tasks are not rescheduled between intervals or cannot be performed in parallel, the workload metric, shown at the bottom, can be thought of as the degree of overload, or the likelihood that one task must be delayed or dropped.

load is calculated as greater than 100 percent, can be identified as potential single channel bottlenecks. Finally, evidence suggests that errors are likely to appear when percent channel occupancy (TR/TA) exceeds 80 percent, (Parks and Boucek 1989), thereby pointing to the 80 percent figure as a plausible red line of workload.

The straightforward assumption of time as the key component for performance prediction and that only a single task can be performed at one time, which is consistent with single-channel theory, appears to be adequate in many circumstances (Hendy, Liao, and Milgram 1997). But there are many more when these assumptions break down, and single-channel theory underestimates the capabilities of human multitasking (Dixon, Wickens, and Chang 2005; Sarno and Wickens 1995). Two of the most important of these are automaticity (reduced effort) and multiple resources. We discuss the first of these below in this chapter and again in chapter 7, while deferring extensive treatment of the second till chapter 8. However, first we consider a second classic historical aspect of attention research closely related to identifying the bottleneck of attention: the auditory shadowing work and its relevance to early versus late selection theory.

Auditory Shadowing: Early versus Late Selection Theory

As noted in the previous section, single-channel theory has been supported by several aspects of the PRP dual-reaction-time paradigm. However, as also noted, basic research addressing the question of where in the sequence of information processing that single-channel bottleneck lies has indicated that much of early sensory and perceptual processing may take place in parallel before the bottleneck. A corresponding issue has been addressed from the perspective of a very different attention paradigm that dominated much of the attention work in the 1950s and 1960s; the study of auditory selective attention through shadowing (e.g., Cherry 1953; Moray 1969; Treisman 1964a, 1964b).

In the shadowing paradigm, experimenters guarantee that the listener's attention is locked onto a given heard channel of usually verbal information by asking the listener to repeat, word for word, the message delivered to the attended ear or attended channel. To the extent that this shadowing is fluent, attention can be inferred to be focused appropriately. While this is going on, other information is presented on another channel (e.g., to the other ear), and investigators examine how much of this other message is processed through a variety of on-line techniques (e.g., brain wave measurements, shadowing disruption) or retrospective techniques (e.g., "Did you notice anything unusual about the unattended message?"). If data suggest that nothing of the unattended message is processed at all, the bottleneck is presumed to be early and, more specifically, prior to stimulus recognition and semantic processing (Broadbent 1958). If there is evidence for understanding of the unattended message, then the bottleneck is presumed to occur after perception and recognition (e.g., Deutch and Deutch 1963; Keele 1973)—that is, late selection. If responses relevant to the unattended message are inadvertently selected (i.e., the unattended message is mistakenly shadowed), then, effectively, there is presumably no bottleneck at all.

Although the earliest shadowing research found evidence to suggest that only physical properties of the unattended message were processed—the bottleneck, or filter, was quite early (Broadbent 1958; Cherry 1953)—three classic studies in the 1960s provided strong indication for semantic processing of the unattended message:

(1) Moray (1959) found that people sometimes noticed the appearance of their own name in the unattended message (although only about a third of the time).
(2) Treisman (1964b) found that people's shadowing of the attended message was disrupted when the unattended message dealt with similar conceptual material—even when that message was in a different language from the attended ear—for bilingual speakers. The language difference is important because it suggests that any interference could not be due to similar sounding words (i.e., physical properties).
(3) Treisman (1964a) found that if the unattended message involved a phrase that was perfectly appropriate in the context of the part of the attended message just heard, people would inadvertently switch to repeat the contextually appropriate—but should be ignored—phrase before realizing their error and quickly switching back.

All three sources of evidence clearly indicate some perceptual and semantic processing of a message that should be, by instructions, shut out of attention. Thus, even though this and a great deal more evidence suggests that some semantic processing can take place without attention (perception takes place before the bottleneck), the emerging evidence also suggested that this processing was much shallower and less effective than that supported by the focus of attention. Snippets of an unattended message for example, might not be remembered more than a few seconds after they were heard. Such finding led Treisman to postulate an attenuation theory in place of a true bottleneck theory, proposing that unattended material is attenuated in its processing rather than either processed at full strength—the view of the true late selection theorists—or not processed at all.

Most recently, however, substantial evidence has accumulated supporting Broadbent's (1958) original bottleneck model (minimal semantic processing of the nonattended channel). Such research reveals that when attention switches do occur—for example, to one's own name (Wood and Cowan 1995)—they are the result of unwanted and uninstructed shifts to the nonattended channel, with a resulting temporary disruption of shadowing the to-be-attended message. Lachter, Forster, and Ruthruff (2004) referred to this as a slippage of the attention filter rather than a leakage of that filter as Treisman's attenuation theory would propose.

The aforementioned findings from the shadowing paradigm are fairly consistent with those from the psychological refractory paradigm by highlighting that some early sensory processing of the nonattended message does occur—as was the case with S_2 in the PRP paradigm and is the case for physical properties of the unattended message in the shadowing paradigm. But this early processing is not extensive and is much less than when attention is focused. Importantly this early processing, whatever its extent and degree, by being preattentive can also be said to be automatic—that is, occurring in the absence of the conscious allocation of attention. Thus, we turn now to the concept of automaticity, examining in particular its status at the opposite end of the capabilities spectrum of human attention from the most severe bottlenecks of single-channel theory.

Automaticity

Over a century ago, William James (1890, p. 57) pondered the span of consciousness, wondering "how many entirely disconnected systems or processes can go on simultaneously." He concluded that "the answer is, *not easily more than one, unless the processes are very habitual*; [automatic] *but then, two, or even three*, without very much oscillation of the attention. Where, however, the processes are less automatic … there must be a rapid oscillation of the mind from one to the next, and no consequent gain of time" (ibid., italics in original). More recently, investigators have observed and documented the gradual improvement in time-sharing (i.e., divided attention) performance as people continuously practice a task (Bahrick and Shelley, 1958; Schneider and Shiffrin 1977) and as the performance of well-trained experts is compared with that of poorly trained novices (Damos 1978). Such changes in performance can be readily accounted for by a simple model that (1) predicts that interference between tasks is a function of the demands of the tasks for a limited supply of mental resources or mental effort (Kahneman 1973), and (2) predicts that the resource demand of a task diminishes with practice and experience until a point of resource-free automaticity is reached. At this point, two tasks can be time shared with no dual-task decrement in either of them. Such is the status that is often attributed to highly learned skills, like walking.

In this regard, it should be noted that the term *automaticity* is a broad one that characterizes several aspects of performance (e.g., consistent, fast, error free), only one of which is perfect time sharing. In particular, the term may also characterize a response that is automatically triggered, cannot be stopped, and therefore may in fact interfere with an ongoing activity. We consider this latter facet of automaticity in our discussion of focused attention in perceptual processing in chapter 6, as well as of task switching and task management in chapter 9. Here we restrict ourselves to the time-sharing implications of automaticity.

Within the last few decades, research has addressed the properties of tasks or their training and practice that can best produce this state of automaticity that avails perfect time sharing (Logan 1990; Schneider and Shiffrin 1977). The seminal work of Walter Schneider, Rich Shiffrin, and Dan Fisk (e.g., Fisk, Ackerman, and Schneider 1987; Schneider 1985; Schneider and Fisk 1982; Schneider and Shiffrin 1977) focused heavily on the essential role of consistency in producing automaticity. In a prototypical task, a small set of letters are designated as targets. The participant is then presented a visual display of one or more letters on each trial and is asked to determine as fast as possible whether or not the display contains a target. When the mapping of stimuli to categories is consistent—that is, when the set of target letters is held constant for many trials—automatic perceptual processing is eventually achieved such that the letter detection task can be performed in parallel with other attention-demanding tasks (Schneider and Fisk 1982) and is as fast with several targets as it is with one. In contrast, extensive practice on the same letter categorization task but with the set of target letters varying from trial to trial produced far less improvement and failed to eliminate a decrement in dual-task performance. This is called controlled processing in contrast to automatic processing. In other words, automaticity does not emerge from practice alone but only from practice with a consistent set of target stimuli.

Fisk and Schneider (1983) generalized this phenomenon to circumstances in which it is the semantic target category that is consistent, not just a particular letter, indicating that automatic detection does not depend only on consistency of physical stimulus. Researchers have also demonstrated consistency-based automaticity with tasks using nonverbal stimuli, like the time–space trajectories of aircraft viewed by an air-traffic controller (Schneider 1985).

Though Schneider and his colleagues have emphasized automaticity of perceptual classification in their research, others have examined the automaticity of both cognitive skills (Logan 1988, 2002), and motor skills (Schmidt 1988; Summers 1989), the latter embodying the concept of a motor program, such as that involved in signing ones own name, skilled typing, or logging into a familiar computer. The underlying theme in all these applications remains that consistency, coupled with practice, yields automaticity.

The applied importance of automaticity theory is realized in at least two areas. The first of these is directly related to training attentional skills, which are discussed in more detail in chapter 11. Certainly trainers of various complex skills are interested in how to bring about automaticity in these skills, or their components, most effectively (Fisk, Ackerman, and Schneider 1987; Schneider 1985). There are two reasons for this. First, components that are automatized (e.g., steering and lane-keeping aspects of driving) will avail ample resources for performance of concurrent tasks that are not ever automatized over many repetitions perhaps because of their lack of consistency (e.g., noticing the appearance of a pedestrian in the roadway). Second, in a complex multicomponent task—such as flying an airplane or directing the basketball offense as a point guard—if it is possible to find components that can be automated, because of their consistency, these make strong candidates to be uncoupled from the full task and submitted to extensive part task training; such a candidate might be dribbling the basketball while moving. Hence, when these components are reintegrated into the complex multitask skill, they will not divert resources that can otherwise be allocated to the complex concurrent subtasks (Lintern and Wickens 1991).

The second application of automaticity theory revisits the timeline modeling approach to workload prediction discussed already. That is, although a true single-channel model predicts that tasks cannot overlap on the timeline—and therefore a workload (TR/TA ratio) of greater than 100 percent mandates a loss in performance of one or another task—a model that incorporates the development of automaticity should enable some time sharing or task overlap. Such assumptions have been embodied in more recent approaches to workload prediction that have incorporated task demand scales that can characterize the relatively low-resource demands of easy tasks, that is, automatized or nearly so (Aldrich et al. 1989; Laughery, LeBiere, and Archer 2006). This issue is considered in more depth in chapter 7. However, it is noted here that, given the challenges of quantifying task demand, strict single-channel assumptions, even if they may not be fully accurate, are often adequate and are preferred in application because time is so much easier to quantify—and to computationally model—than is the degree of automaticity or the effort demanded by the task (Liao and Moray 1993). Models without such an effort component can often do an adequate job of predicting multiple-task performance (Sarno and Wickens 1995).

Conclusions

In conclusion, we have examined two historical concepts—single-channel theory and automaticity theory—that are both intuitive, representing opposite extremes of human attentional capabilities, each quite evident in real-world tasks. The following chapters show each of these concepts reemerging in different forms, characterizing different aspects of attention. For example, in chapter 5 on visual search, we see contrasted clearly the single-channel theory assumptions of serial search, with the automaticity concepts of parallel search. Chapter 6 discusses the automaticity with which different features of a single object are processed. Chapter 7 discusses in detail the mediating concept of effort, or resource demands, that can create single-channel theory behavior when it is high, automatic behavior when effort is low, and something in between when effort demands are moderate.

3

Attention Control

Introduction

Many real-world tasks—driving, flying, process management—require an operator to monitor multiple information sources over an extended period of time. In such cases, as is discussed in chapter 4, the operator typically learns to allocate attention in a way that is adaptive, shifting attention between various channels with a frequency determined by their relative importance and bandwidth. Individual shifts of attention, however, occur over the course of mere seconds or less. Moreover, even when the operator's attentional scanning strategies are well tuned over the long run, it may still be useful to alert the operator to unexpected, infrequent, or high-priority events, interrupting the normal path of attentional scanning to ensure that important information is quickly encoded. Even an experienced and attentive driver, for example, might benefit from an alert that announces when the fuel gauge is approaching empty, from an alarm that signals an impending side collision, or from the bright reflective clothes of a nighttime jogger. A designer may sometimes also anticipate users who bring little if any top-down knowledge-driven guidance to bear on their interactions with a product or system. Effective design in such instances requires cues to guide attention in a bottom-up manner toward crucial information. When designing instructional graphics or animations, for example, an educator may need to cue attention to important details that students, unfamiliar with the material, might otherwise overlook (Lowe 2003).

In such cases, the system designer is faced with a challenge of attentional control—of catching the operator's attention and orienting it toward a timely piece of task-relevant information. Although the problem of cuing attention to an information channel seems straightforward, the designer may face difficulty in creating a cue that is noticeable but not disruptive. A visual cue that is readily detectable in a sparse or static display, for example, may be lost in a cluttered bank of blinking lights, spinning dials, and flickering monitors, particularly if it is visible only in the visual periphery or if the operator is engaged in another task (Nikolic, Orr, and Sarter 2004). Conversely, a cue that is conspicuous or persistent enough to guarantee detection in a busy environment may be overly distracting and subjectively annoying (Bartram, Ware, and Calvert 2003). The present chapter discusses several basic issues of attention capture and capture failures before turning to a discussion of the most important applications of attention control: the role of automation in directing attention to critical events via alarms and alerts.

Inattentional and Change Blindness

The issue of attentional control can be studied from the perspective of what captures attention. However, the issue is also relevant in cases where environmental events fail to draw attention, as in the phenomena of inattentional blindness (Carpenter 2002; Mack and Rock 1998) and change blindness (Rensink 2002; Simons and Levin 1997). Inattentional blindness, sometimes described as the *looked-but-failed-to-see* effect (Herslund and Jørgensen 2003), occurs when a lapse of attention causes an observer to overlook an object that is prominent in the visual field and is well above sensory threshold. Mack and Rock (1998), for instance, found that when subjects focused their attention intently on a shape in their central vision, they often failed to detect an unexpected probe object that appeared at the same time in the visual periphery. More surprisingly, if attention was directed to the visual periphery, many subjects failed to notice the unexpected probe even when it was presented in their central vision directly where their eyes were fixated. These effects were not simply the result of poor stimulus visibility, it is important to note, as the probes were easily detected when subjects expected them. Failures to notice the probes were attentional, not sensory. Looked-but-failed-to-see errors are easy to produce and sometimes startling in their strength (e.g., Simons and Chabris 1999) and have been implicated as a common cause of traffic accidents (Herslund and Jørgensen 2003).

Change blindness, another form of visual lapse, occurs when an observer fails to detect an event (e.g., discrete change) in the environment around him. Thus, whereas inattentional blindness is a failure to notice something here and now, change blindness is a failure to notice that something is different from what it was. For an online chat box user sitting back down at his computer, this could be a failure to notice that a new message has arrived; for a driver turning her eyes back to the road after a glance at the radio dial, it might be a failure to notice that the vehicle ahead now has its brake lights on (Pringle et al. 2001); for a pilot contending with the demands of a busy cockpit, it may be in a failure to notice the onset of a green light in the cockpit display that indicates a change in flight mode (Sarter, Mumaw, and Wickens 2007). In a face-to-face conversation, more remarkably, it can even mean that a speaker fails to notice that the person to whom he was speaking has disappeared and been replaced by someone new. In a study by Simons and Levin (1998), an experimenter approached pedestrians on a college campus to ask for directions. As the experimenter and a pedestrian talked, their conversation was interrupted by a pair of workers (confederates of the experimenter, in actuality) who barged between them carrying a wooden door. As the workers tromped through, the original experimenter snuck away behind the door, and a new experimenter was left standing in his place. In most cases, the conversation carried on normally, with the pedestrian failing to notice that he was speaking to someone new.

How do such lapses occur? Very often, CB is a failure of attention. Data indicate that changes are likely to go unnoticed if they are not attended when they occur. Under most circumstances, a change in the visual environment produces a transient, a brief flutter of motion or flicker. These signals tend to attract attention and help ensure that the change is noticed. However, the change will often go undetected if it occurs while the observer is looking away, or if its accompanying transient happens to be hidden by an occluding object (Simons and Levin 1998), the flicker of a display (Rensink, O'Regan, and Clark 1997), egomotion (Wallis and Bülthoff 2000), other transients produced by concurrent events (O'Regan, Rensink, and Clark 1999), or even just an eye movement (Grimes 1996) or blink (O'Regan et al. 2000). In the terminology of Rensink (2002), a dynamic change, or a change in progress, is generally easy to notice, but a completed change, one that occurred while attention was elsewhere or vision is obscured, is not.

It is easy to envision real-world circumstances in which the difficulty of detecting completed changes may hinder task performance. Consider an operator monitoring a large dynamic display of air traffic. At any given time, most of the display will fall outside the momentary focus of attention. As a result, a large number of changes will be likely to escape notice. Indeed, in one simulation pilots monitoring such a traffic display detected fewer than half of such changes, including those such as changes in course and altitude of another aircraft with important implications for safety (Muthard and Wickens 2002). When display monitoring was coupled with an attention-demanding flight control task, change detection rates dropped to well under 25 percent (Steltzer and Wickens 2006).

Interestingly, recent work has indicated that despite their remarkable blindness to completed changes, people incorrectly tend to assume that such changes will be noticed easily, a phenomenon that has been termed *change blindness blindness* (Levin et al. 2000). Thus, not only are operators performing a complex task likely to overlook much of the activity going on around them, but they are also likely to markedly overestimate the degree to which their knowledge of the world is current, a metacognitive error that may discourage them from seeking new information when appropriate. Theorists argue that the first step toward good situation awareness is to notice objects and events in the surrounding environment (Endsley 1995). The phenomena of IB and CB make clear that attention plays an indispensable role in this process (Wickens and Rose 2001).

Covert versus Overt Attention

How is visual attention allocated? Helmholtz (1867, 1962), describing an experiment in which he tested his ability to see images briefly illuminated by a spark of light, noted that the direction of visual attention was not strictly locked to the direction of the eyes. Rather, he observed, it was possible to "keep our attention … turned to any particular portion we please" of the visual field, "so as then, when the spark comes, to receive an impression only from such parts of the picture as lie in this region" (1925, p. 455) even while the eyes were fixed elsewhere. James (1890, p. 434) likewise noted a difference between "the accommodation or adjustment of the sensory organs" and "the anticipatory preparation of from within of the ideational centres concerned with the object to which attention is paid." Following Posner (1980), this difference is now described as the distinction between overt and covert visual orienting. An overt attention shift is one that involves a movement of the eyes, head, or body to bring the sensory surface into better alignment with the object of interest. In vision, this is most often a saccade: an abrupt, rapid movement with which the eyes dart from one point of fixation to another. A covert attention shift, conversely, is a change of mental focus in the absence of any physical movements. Thus, a computer user may orient overtly by moving her eyes to look directly at a text message that has just popped up on her screen, or she may notice the message covertly in her peripheral vision without turning to look directly at it. Likewise, a person in a noisy room may overtly turn his head and lean forward to listen more carefully to what a friend is saying, but may covertly listen more intently with no outer sign of effort while trying to eavesdrop on someone else's conversation.

To separate the effects of attentional processes from the effects of visual acuity, basic attention researchers often ask their subjects to make judgments of objects in the retinal periphery while holding the eyes fixed on a central point. In most real-world tasks, however, eye movements are common. In viewing a naturalistic scene, observers generally make two to three eye movements per second, pausing in between for fixations of 250–300 ms[*] (Rayner 1998). Indeed, Moray and Rotenberg (1989, p. 1320) speculated that covert attention shifts are of little consequence in most applied contexts—that "the more a task is like a real industrial task, where the operators can move their head, eyes and bodies, can slouch or sit up, and can, in general, relate physically to the task in any way they please, the more 'attention' is limited by 'coarse' mechanisms such as eye … movements." Nonetheless, information gathered from the study of covert attention is valuable to the applied attention researcher. Research has found that in vision, overt and covert movements are subserved by many common neural mechanisms (Corbetta 1998) and are functionally linked, though in an asymmetrical manner; it is possible to move attention covertly without also moving the eyes, but a saccade cannot be executed until covert attention has shifted to the target location (Deubel and Schneider 1996; Hoffman and

---

[*] Saccades themselves last only a few tens of milliseconds, and visual input is processed only during the intervening fixations. Visual processing is suspended while the eyes are in flight, leaving the viewer effectively blind for a fraction of a second. This effect, known as *saccadic suppression*, ensures that the viewer does not experience a smeared retinal image when a saccade occurs.

Subramaniam 1995; Kowler et al. 1995). Stimulus attributes that attract covert visual attention therefore tend to have similar, though less powerful, effects on saccades (Irwin et al. 2000; Theeuwes et al. 1998). Understanding the workings of covert attention tells us much about overt attention.

The Spotlight of Visual Attention

Covert visual attention is often described as analogous to a spotlight (Cave and Bichot 1999), enabling selective processing of the "illuminated" region of the visual field. While the spotlight model—and related conceptualizations of attention as a mental zoom lens (Eriksen and St. James 1986) or gradient of resources (Downing and Pinker 1985; Laberge and Brown 1989)—carries many implications that are at best unconfirmed (Cave and Bichot 1999) and in some cases have been disproven (e.g., Duncan 1984; Kramer and Hahn 1995), it remains popular and captures some aspects of selective visual processing.

Evidence for the spotlight model originated in the work of C. W. Eriksen and colleagues (Eriksen and Hoffman, 1972, 1973) using what has become known as the flanker paradigm. In a typical flanker experiment, the subject's job is to identify a target letter appearing at a particular location. Commonly, two or more target letters are linked to the same response. For example, the subject may be asked to press one button if the target is an H or an M and to press a different button if the target is an A or a U. The target is surrounded on either side by task-irrelevant flanker letters that are mapped to the same or a different response as the target. Subjects are asked to ignore the flankers and to respond solely on the identity of the target. Despite these instructions, the classical finding in the flanker paradigm is that response times (RTs) are influenced by the identity of the flanking stimuli, being longer when the flankers are response incompatible with the target than when they are response compatible. These effects, moreover, are strongest when the separation between the target flankers is small and often disappear when the separation between target and flankers exceeds approximately one degree of visual angle (Eriksen and Hoffman 1973; Yantis and Johnston 1990). Thus, flankers very near the target item appear to be processed to the point of recognition even when the subject wishes to ignore them, a failure of focused attention selectivity. Eriksen and Hoffman (1973) took their findings as evidence of a spatial attentional focus with a minimum size of about one degree. As is discussed more fully in chapter 6, this conclusion bears an important implication for display design: Displays and maps designed with items too closely together—even if they do not actually overlap—will hinder the focus of attention on any given item.

Posner's Cuing Paradigm

In the real world, cues to shift attention may sometimes be inappropriate or just plain wrong. This issue was captured in an experimental procedure

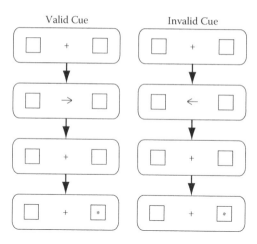

Figure 3.1 Sample trials from a Posner cuing task. After a brief get-ready interval at the beginning of the trial (top panel), a cue is presented to direct attention to one of the potential target locations (second panel). In this case, the cue is an arrow that appears in the center of the display, pointing to one the two boxes onscreen. After a brief delay (third panel), a target object appears at one of two or more potentially locations. On valid cued trials, the target appears in the cued location (left column). On invalid cued trials, it appears in a location that was not cued (right column).

known as the cuing paradigm, developed by Posner and colleagues (Posner, 1980; Posner, Snyder, and Davidson, 1980). In a typical version of Posner's cuing change to procedure the observer is asked to keep her eyes on a central fixation mark and to make a speeded detection judgment of a target signal that can appear in the visual periphery on either side. The observer's attention is manipulated by a cue that appears prior to signal onset. The columns of Figure 3.1 illustrate the events within two different types of experimental trials. Here, the cue is an arrow pointing toward one of the possible target locations. On valid cue trials, as illustrated in the left column, the target signal appears at the cued location. On invalid cue trials, as illustrated in the right column, the target signal appears at an uncued location. Generally, cue validity is above chance, such that target is more likely to appear at the cued than at the uncued location. The effects of attention are measured by comparing RTs for target detection following cues of different validity; an attentional cost obtains when RTs for invalid cue trials are longer than for control trials on which no location is cued, and an attention benefit obtains when RTs for valid cue trials are shorter than for control trials. Studies have confirmed that even for a task as simple as detecting the onset of an above-threshold spot of light, valid cues reduce RTs and invalid cues increase them (e.g., Posner, Snyder, and Davidson 1980). Similar effects obtain when a discrimination task is used instead of a detection task (e.g., ibid.), and when judgment accuracy is used as the dependent variable (e.g., Lu and Dosher 1998).

Posner, Snyder, and Davidson (1980) suggested that cues improve signal detection and discrimination by allowing subjects to align the mental spotlight of attention with the expected target location. What effect, more exactly, does this attentional orienting have? Although cuing costs and benefits are observed in simple uncluttered displays (e.g., Carrasco, Penpeci-Talgar, and Eckstein 2000), suggesting that cues can enhance the perceptual signal strength, data show that the benefits of attentional cuing are substantially larger when the target object is surrounded by easily confusable distractor objects (Shiu and Pashler 1994) or embedded in visual noise (Lu and Dosher 1998) than when it appears by itself against an open background. In real-world environments, these costs of invalid cueing can be quite large, in the order of several seconds (Yeh, Merlo, Wickens, and Brandenburg 2003). These results suggest that one effect of cuing is noise exclusion: the ability to ignore distracting or target degrading information.

Central versus Peripheral Cuing

As might be expected, characteristics of the cue stimulus itself also influence attentional performance. This was first demonstrated by Jonides (1981), who compared the effects of central and peripheral cues. A central, or endogenous, cue denotes the cued location symbolically but does not actually appear at that location. Typically, it is presented at a fixation point in the center of the visual field, though it need not always do so. A central cue may be an arrow, as in Figure 3.1, a word, or any other kind of symbol, so long as it does not directly mark the location to which attention should be shifted. A peripheral, or exogenous, cue, conversely, is a transient signal—usually a luminance change or burst of motion—that appears at the location to which attention is to be shifted. In Figure 3.1, for example, attention could have been exogenously cued to the target location by blinking or briefly brightening the box at the to-be-attended location. Jonides found that such peripheral cues, unlike central cues, tend to draw attention even if they do not predict the target location with accuracy better than chance, and tend to be effective even under relatively high levels of cognitive workload. For these reasons, peripheral cues are said to elicit reflexive attention shifts, whereas central cues are said to require voluntary orienting.

Other work in this area has found that reflexive attentional shifts are faster than voluntary shifts (Müller and Rabbitt 1989) but that the attentional benefits of reflexive orienting fade quickly if they are not maintained by voluntary processes. As such, reflexive visual attention shifts are sometimes described as transient and voluntary shifts as sustained (Nakayama and Mackeben 1989). As a real-world analog, we might consider a head-up display (HUD) that cues the driver's attention to hazards in the roadway. Such a cue could be either a salient light on the windshield along the line of sight to the hazard (corresponding to a peripheral cue) or an arrow in the center of the driver's forward view, pointing toward the hazard (central cue). Consistent with the data from basic research, Yeh, Wickens, and Brandenburg (2003) obtained results suggesting that peripheral cues are more effective at least so long as they are reliable and do not obscure the actual target; these issues are addressed at the end of this chapter.

As the previous example suggests, the distinction between exogenous and endogenous orienting can help display designers know how to create signals or cues that most effectively guide attention. In light of this, researchers have invested great effort to determine precisely what stimulus properties are capable of eliciting reflexive attention shifts. Recent findings have in fact begun to blur the line between voluntary and reflexive processes. Contrary to Jonides's (1981) early findings, a number of researchers have documented instances in which symbolic cues that are highly overlearned or otherwise very easy to interpret seem to elicit reflexive attentional movements. Familiar words such as *left* and *right,* for instance, or social cues such as a face in which the eyes are looking in a particular direction, may produce reflexive attentional orienting even if they are not predictive of the likely target location (e.g., Gibson and Kingstone 2006).

Nonetheless, some conclusions about the relative power of various cues to grab attention—of interest both to researchers and designer—can be drawn. At least three factors help determine the likelihood with which a visual signal captures and holds attention. The first is stimulus salience, where salience is a signal-to-noise measure of the feature contrast (e.g., color, motion, luminance contrast) between the target and the surrounding stimuli relative to the feature variability among the surrounding (Itti and Koch 2000; Nothdurft 1992; Theeuwes 1994). Motion and color contrast, which provide strong cues for object segmentation in the natural visual environment (Gibson, 1979; Regan, 2000), may be particularly salient.

Thus, for instance, a tilted gauge among a display of vertical gauges can attract attention readily, whereas a tilted gauge within a display of randomly oriented gauges may not. These issues are discussed more in the next two chapters.

The second factor that improves attention capture is stimulus newness. In Jonides's (1981) original study of endogenous and exogenous orienting, as noted, peripheral cues were arrows that flashed on-screen briefly at the cued location. Evidence has since suggested that such abrupt onsets are especially potent in capturing visual attention. More specifically, studies by Yantis and colleagues (Jonides and Yantis 1988; Yantis and Hillstrom 1994; Yantis and Jonides 1984, 1990) produced evidence that the appearance of something new within a scene or display triggers an attention shift. Under this hypothesis, it is the status of a stimulus as something new that captures attention rather than simply a luminance change or other visual transient. Consistent with this possibility, data indicate that a luminance change by itself is relatively less likely to capture attention if it does not signal the appearance of a new object (Irwin et al. 2000; Jonides and Yantis 1988). Echoing these findings from the basic research lab, a study of battlefield display monitoring found that completed changes indicating the appearance of an enemy unit were reported at an 84 percent rate, whereas those involving the removal of a unit were only detected 44 percent of the time (Wickens, Thomas, and Young 2000).

The third factor modulating the likelihood of capture is the observer's attentional set. Any visual signal, no matter how salient, is more likely to attract and hold attention if the observer is searching for it or is otherwise prepared to let it grab attention (Folk, Remington, and Johnston 1992; Most and Astur 2007). The ability of a signal to grab and hold attention is therefore strongly modulated by its validity—the degree to which it predicts the occurrence and location of an impending target stimulus (Folk, Remington, and Johnston 1992; Yantis and Egeth 1999). The importance of attentional set in applied domains is obvious in the next section, which discusses the effects of alert and alarm reliability.

Auditory and Cross-Modal Cuing

Although attentional control has been studied most intensely in the visual modality, researchers have adapted Posner's (1990) cuing task to explore auditory and tactile attentional processes as well. In a series of experiments by Spence and Driver (1994), for example, subjects were asked to make speeded judgments of auditory signals that could originate from either the left or right of the subjects' midline. The target signal on each trial was preceded by an auditory cue that came from either the same or the opposite side at varying intervals prior to target onset. In some experiments the cue was unpredictive of the target stimulus location (i.e., the target was equally likely to appear on the cued and uncued sides), and in other cases it was predictive with 75 percent validity. Consistent with data from studies of visual attention, judgments were faster when the target stimulus originated on the cued side than when it came from the uncued side. The effects of auditory cuing were qualified in two important ways, however. First, cuing benefits were larger and longer lasting when auditory cues were predictive of the target side than when they were unpredictive. Thus, as in vision, exogenous attentional processes were most effective when aided by endogenous processes. Second, cues speeded auditory discrimination judgments (e.g., Was the pitch high or low?) but had no effect on the speed of auditory detection. The reason for this, the authors speculated, was that target detection did not require focused auditory attention information. This finding points to the particular value of auditory stimuli as alerts or alarms in applied settings.

An additional characteristic that makes auditory stimuli useful as alarms is that they summon not only auditory attention but also visual attention. Indeed, data suggest that visual, auditory, and even tactile attention are all spatially linked (Driver and Spence 2004; Spence, McDonald, and Driver 2004), such that attentional cuing in one perceptual modality can produce RT benefits for targets in a different modality. A cue that directs auditory

attention to one side of the observer's midline, for example, will speed judgments of visual targets on that side as well; conversely, a cue that directs
visual attention to one side of the visual field will also facilitate judgments of
auditory stimuli from that side (Spence and Driver 1996). Auditory and visual
attentional also appear to travel together in tasks that required sustained
monitoring of information channels. For example, a study by Spence and
Read (2003) asked subjects to monitor a stream of speech while they performed a simulated driving task and found that performance was better
when the auditory stream came from a speaker directly in front of the driver,
near the focus of visual attention, than when it came from a speaker to the
driver's side. Spatial links are also seen between visual and tactile attention
(e.g., cuing visual attention to a position near one of the hands facilitates
judgments of vibrotactile targets with the same hand) and between auditory
and tactile attention (Lloyd, Merat, McGlone, and Spence 2003; Sarter 2007;
Spence, Pavani, and Driver 2000).

Interestingly, cross-modal cuing effects do not always require that display
channels in different modalities be physically coincident. Rather, it is sufficient that there be a mental correspondence between locations. For example, an auditory or tactile cue that comes from behind a driver facilitates
the detection of an impending rear collision more than one that originates
from the front—even if the only way for the driver to detect the collision is
by checking the rearview mirror in the forward field of view. This is true
even if the attentional cues are purely exogenous. In other words, drivers'
reflexive attention-shifting processes recognize the well-learned spatial correspondence between the location of the nonvisual cue behind them and the
visual information depicted in the rearview mirror (Ho and Spence 2005; Ho,
Tan, and Spence 2005).

Two additional characteristics of the spatial links among auditory, visual,
and tactile attention are also noteworthy to the engineering psychologist
(Spence and also Driver 1997). First, these links are not simply the results
of the observer's task strategy but are obligatory. Thus, even when it might
be beneficial to divide auditory, visual, or tactile attention between different locations—either to monitor spatially separated streams of information (Driver and Spence 1994) or to filter away an irrelevant stream in one
modality—people find it difficult do so (Spence and Walton 2005; Spence,
Ranson, and Driver 2000). Second, though cross-modal interactions are
seen even with purely covert attention shifts (Driver and Spence 2004), they
also affect eye movements. For example, observers are capable of targeting
saccades toward a sound source quite accurately (Frens and Van Opstal
1995) and are speeded in executing a saccade when a sound or vibrotactile
stimulus coincides with the visual saccade target (Amlôt, Walker, Spence,
and Driver 2003).

The implications of these facts for display design are straightforward.
When displays in multiple perceptual modalities present information that
is to be compared or integrated—for example, when a speech stream is used

to provide directions to a driver whose vehicle is displayed on a dashboard-mounted electronic map, or when an auditory alert is used to cue a pilot's visual attention to a display indicating a system error—the display channels in the various modalities should be arranged near one another. Conversely, when an operator is expected to attend to a single channel while filtering away distracting information in another channel, the to-be-attended channel should be spatially separated from the distractor channel even if the two channels are in different perceptual modalities. Further implications of attentional control for display are discussed in the next section.

Applied Implications of Attention Capture Research: Alarms and Attention Guidance

Designers have used what is known about attention control to create displays that capture and then direct attention to important events and information channels in real-world systems. Of course, advertisers have long known about many of these factors, taking advantage of them to draw attention toward products on shelves, in the Yellow Pages, and as pop-up items on Web pages. The following sections discuss two sorts of applications: spatial cuing systems and alarms and alerts (Pritchett 2001; Stanton 1994; Wogalter and Laughery 2006). Both have in common the goal of capturing attention, often in a busy multitask environment, though they differ in that alarms and alerts tend to direct attention implicitly (e.g., the fire alarm will lead a person to attend to sounds or smells in the hallway), whereas attention guidance systems, by definition, do so explicitly. Both are forms of automation in which a computer system decides that the human should redirect attention. As such, both have the potential to err and, in the language of signal detection theory, to do so in either of two different ways: (1) by committing a false alarm, directing attention to a problem that does not exist (Breznitz 1983); or (2) by committing a miss, failing to direct attention to a problem that is real. More regarding the issue of highlighting to achieve attentional direction is discussed in chapter 5.

A framework for understanding these systems is provided in Figure 3.2, which depicts the automation system (top) and the human (bottom) monitoring in parallel a domain of raw data in which events—typically hazards—occur (Getty et al. 1995; Sorkin and Woods 1985). Performance of both the automation system and the human perceiving automation's output in parallel with the raw data and integrating the two can be represented in the context of the signal detection theory matrix, shown at the top of the figure. Such a matrix considers the two types of responses—signal present or absent—that can be made in the face of the two states of the world (signal or no signal), and the matrix then yields the four types of decision events that can result.

Following is discussed the research bearing on four key factors that influence the effectiveness of such systems in appropriately capturing and guiding

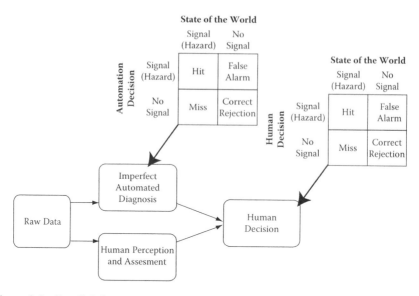

Figure 3.2 Parallel human and automation alerting system. The classic signal detection theory matrix is shown in the insets at the top, representing performance of the automated alert system alone and of the human relying both on the automated system and direct perception of the raw data. Later in the chapter, a discussion is given of the research bearing on four key factors that influence the effectiveness of such systems in appropriately capturing and guiding attention: (1) physical salience (bottom-up conspicuity); (2) reliability; (2) response bias; and (4) the frequency of the critical events that occur in the raw data of the alerted domain.

attention: (1) physical salience; (2) reliability; (3) response bias; and (4) the frequency of the critical events that occur in the raw data of the alerted domain.

Alert Salience

Drawing from consideration of the fundamental properties of the human senses (Proctor and Proctor 2006), it becomes apparent that auditory stimuli, or auditory onsets, are the most reliable attention grabbers for alarm systems. This is because the detection of auditory events is omnidirectional. That is, a sound is nearly equally salient no matter how the head is oriented. This of course is not the case with vision, where salience falls off rapidly with visual angle away from the fovea and visual events tend to be invisible at angles beyond around 90 degrees. Recently research has also focused on the tactile modality as being nearly as efficient and omnidirectional as the auditory modality in capturing attention (Sarter 2007; Sklar and Sarter 1999; van Erp 2006; van Erp, Veltman, and van Vern 2003).

Despite the auditory superiority in direct alerting, auditory alerts can also present problems because of that very attribute of intrusiveness. They may well interrupt ongoing tasks that could be of even higher priority (Ho, Tan, and Spence 2004; Wickens and Colcombe 2007a,b), a phenomenon known as auditory preemption that is discussed again in chapters 8 and 9. It is in part for this reason that designers have often considered less disruptive visual alerts, particularly for those events that may be less critical for safety (Latorella 1996). In this regard we can again draw on guidance from more basic research.

The most direct generalizations from basic research on attention capture to alarm and alert design discussed already is that onsets tend to capture attention. Hence, the most effective visual alarm will be the flashing or blinking signal, because the flashing entails repeated onsets, any of which may eventually be noticed as the eyes are busy scanning the environment (Wickens and Rose 2001). A second generalization is that unique colors can be effective as alarms and alerts (e.g., a red light in a swarm of green). However, the ability of a singleton to capture attention will be reduced if nearby stimuli are also color coded; as discussed in Chapter 5, the salience of a uniquely colored item is decreased by background stimuli of heterogeneous color.

The most severe limitation to generalization of the conclusions drawn from basic attention capture research is the fact that in most environments where warnings are presented, the visual space that the operator is required to monitor is much larger than the typical computer screen—where, for example, onsets are sometimes guaranteed to capture attention (Yantis 1993). In this light, we must acknowledge that there are many circumstances in which singletons and even onsets do not invariably capture attention. A perfect example here is the failure of the green box onset in automated aircraft to capture the pilot's attention to changes in automation mode (Nikolic, Orr, and Sarter 2004; Sarter, Mumaw, and Wickens 2007). Because important alarms are typically auditory, considerable effort has been taken to define the physical parameters that can convey different levels of urgency in attention capture (Edworthy, Loxley, and Dennis 1991; Marshall, Lee, and Austria 2007). Here, however, caution must be taken because increasing urgency may also lead to increasing listener annoyance (Marshall, Lee, and Austria 2007)—a real problem if the alarms are false, as discussed following.

One final concern when designing visual alerts for attention guidance is the possibility that highly salient exogenous cues may mask the raw data beneath because of their close spatial proximity to those data (i.e., less than one degree). For example, consider the soldier whose attention is cued by the onset of a box surrounding the location of a potential enemy target on a head mounted display (Maltz and Shinar 2003; Yeh and Wickens 2001a; Yeh et al. 2003; Yeh, Wickens, and Seagull 1999). Here, if the raw stimulus information (i.e., the image of the enemy target) is faint, the soldier may not be able to easily discern whether or not a target is actually present behind the visual cue or, worse yet, whether the target is actually an enemy. If the automation is always correct in designating location and identity, then this perceptual challenge will present no problem. But if the automation is less than perfectly reliable in making its classification, the consequences of degrading the view of the raw data image can be severe. Good human factors design would therefore suggest that the concern of masking can be addressed by using an arrow adjacent to and pointing toward the target rather than a shape surrounding the target object. This leads to the second important feature of attention-guiding automation: its reliability.

Alert Reliability

When alert systems are asked to detect dangerous events based on uncertain data or are asked to predict events (e.g., mid-air collision, a hurricane track intersecting a city) in a world that is inherently probabilistic, they will sometimes be wrong, driving their reliability below the perfect value of 1.0. In particular, with predictive alert systems, the longer the look-ahead time or span of prediction, the lower that reliability will typically be (Thomas and Rantanen 2006). Automation reliability, mimicking the concept of cue validity previously discussed (Posner, Snyder, and Davidson 1980) will affect overall system performance as well as two related but distinct attributes of user cognition and behavior: trust and dependence (Lee and Moray 1994; Lee and See 2004; Parasuraman and Riley 1997). Trust represents a subjective belief that automation will perform as expected. Dependence, often correlated with trust, is the actual behavioral tendency of the human to do what automation prescribes (e.g., look where the automation says to look). Clearly, both of these measures will drop as automation reliability declines. But it is easy to envision systems when trust may be fairly low (i.e., the human expects automation to make errors) but when dependence nevertheless remains high, either because (1) the automation, though imperfect, is adequate, or (2) the human operator's workload is high, so attentional resources are needed elsewhere. Because the major attentional implications of imperfect automation are related to behavioral dependence rather than trust, the focus here is on the former—behavioral measure.

It is clear that the total human–system performance of the parallel system shown in Figure 3.2 will improve as diagnostic automation reliability approaches 1.0. However, the relationship between automation reliability and system performance is not perfect. When the automation alone is better than the human alone, combined system performance will generally be better than unaided human performance but worse than the automation by itself (Parasuraman 1987; Wickens and Dixon 2007). Thus, the human typically depends somewhat, but not fully, on automation. However, there is some evidence that when reliability of the automation is very low (e.g., less than around 0.70), the human may continue to partially depend on it even when his or her own unaided performance is superior to that of the automation. Wickens and Dixon analogized automation in such circumstances to a concrete life preserver: the human would be better off letting go of the automation rather than clinging to it. Recently, for example, researchers have found that physicians depending on automated devices to detect tumors

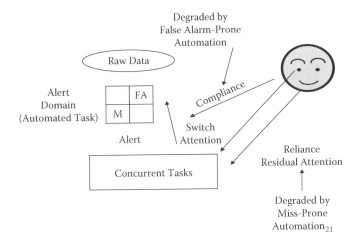

Figure 3.3 Monitoring imperfect automation within a dual-task context.

on mammograms do no better—and in some respects worse (more false alarms)—than physicians without such aids (Fenton et al. 2007). Of course, the other side of this story is that an automated attention guidance system will generally assist human performance, even if it is less than fully reliable, so long as that reliability is above about 0.80 (Dixon and Wickens 2006; Xu, Wickens, and Rantanen 2007).

Automation reliability has particularly important effects on attention within the multitask contexts, in which automation (e.g., a warning or alert) is likely to be most valuable, because attention is usually needed for purposes other than monitoring the raw data (Figure 3.3). A quantitative analysis of the literature indicates that, on the one hand, dependence on automation is greatest when workload is high, such as in a dual-task context (Wickens and Dixon 2007). On the other hand, this literature suggests that it is the automation-supported detection task rather than the concurrent tasks that suffers most as automation reliability declines, as if the human continues dedicating necessary attention to the concurrent task to preserve its performance—an issue of resource allocation discussed in chapters 7 and 8.

In studying the effects of automation detection and diagnosis imperfection, it is always important to distinguish between the effect of a first failure, when the operator encounters the initial failure of a system that has heretofore in his or her experience performed perfectly, and the effect of subsequent failures, which occur at a long-term rate inversely proportional to the automation's reliability level (Molloy and Parasuraman 1996; Yeh and Wickens 2001a; Yeh et al. 2003). The consequences of a first failure are often large and potentially catastrophic, as the operator may have become complacent, assuming a perfect system and therefore allocating minimal, if any, attention to monitoring the raw data (Parasuraman, Molloy, and Singh 1993). Subsequent performance, however, becomes more stable as the operator adjusts his or her attentional strategies to the long-term statistical quality of the system. The next section turns to these.

The Nature of Automation Errors: Error Salience and the Alert Threshold

When automation reliability declines, automation errors become more common. It turns out that it is not only the number of errors but also the kind of errors that influence human trust and dependence. One particularly noteworthy factor that bears directly on attention is the obviousness or salience of the errors that automation does make. If the automated system's errors are made on judgments that seem easy to the operator (Madhavan, Weigmann, and Lacson 2006) or if the consequence of such errors is large, those errors will be more obvious to the operator and hence will undermine trust and dependence to a greater extent.

Whether errors are obvious or subtle, given an imperfect alerting system alarm designers—and often users—typically have the option of adjusting the threshold of the system to trade off between the two kinds of automation errors within the context of signal detection theory shown in Figures 3.2 and 3.3, choosing anywhere between a low threshold (i.e., hair trigger), which ensures that misses are rare but also allows frequent false alarms, and a high threshold, which reduces the risks of false alarms by increasing the probability of misses. In many predictive alerts, such as collision warnings, such misses are actually late alerts (Maltz and Shinar 2004). In the terminology of signal detection theory, this is an adjustment of the alarm's response criterion (beta). In most real world systems, the costs of misses outweigh those of false alarms, and so the threshold is set low (Getty et al. 1995); consider the difference in consequences of a fire alarm that fails to detect a true fire versus one that produces a false alarm.

It turns out that this threshold adjustment will influence two different cognitive states of the human user—reliance and compliance—which have direct impacts on attention and via this exert influences on performance (Dixon and Wickens 2006; Maltz and Shinar 2003; Meyer 2001, 2004), as depicted in Figure 3.3. Reliance is the cognitive state that allows the operator to feel confident that there really is no hazard at the times when the alert is silent. Compliance, conversely, is the cognitive state that allows the operator to act confidently in response to an alarm when it occurs. Both of these states can be thought of as subcategories of automation dependence discussed already. The system, prone to a high miss rate, resulting from setting a high alert threshold, will lower the human's reliance on the automation to issue an alert if there really is a dangerous event. As a consequence, as shown in Figure 3.3, the human will be more likely to reallocate attentional resources or residual attention away from the concurrent tasks to monitor the raw data of the automated task to pick up the events that automation misses. As a result of this allocation, concurrent tasks will suffer (Wickens and Colcombe 2007b; Dixon and Wickens 2006), but the (now frequent) failures of automation to detect those events will actually be easily caught by the operator, who expects those automation misses to occur. We may think of this as a drop in complacency—a healthy skepticism (Maltz and Shinar 2003). In contrast, as shown at the top of Figure 3.3, the false alarm (FA)-prone system, resulting from a low alert threshold, will induce a loss of compliance. Suspecting a false alarm, the human may delay taking an action in response to the sounding of an alarm—whether true or false—or may even decline to respond at all. This is the cry wolf effect (Breznitz 1983; Sorkin 1989). As shown in Figure 3.3, in terms of attentional effects, low compliance reflects a slowing of on the speed of attentional switching to any alarm, whether true or false.

It is noteworthy that even though the states of reliance and compliance are logically independent, in practice they often appear to vary in tandem (Dixon and Wickens 2006). Thus, although a low threshold should leave the human confident that all true events will be alerted—and hence should allow full attention allocation to concurrent tasks—FA-prone systems in fact are often as disruptive of concurrent tasks as miss-prone systems, if not more so. This disruption is seen both in the laboratory (Dixon, Wickens, and McCarley 2007) and in the real world of alerting systems (Bliss 2003). Two explanations suggest themselves. First, as noted, every alert will usually be responded to in some fashion by the operator, whether true or false, and the added frequency of false alerts will thereby cause an added frequency of attention shifts away from the concurrent tasks. This effect alone, however, cannot fully account for the detrimental effects of automation false alarms on concurrent task performance, since this performance suffers even on trials where the automation commits no false alarm (Dixon, Wickens, and McCarley 2007).

Second, false alerts are often more perceptually obvious and cognitively disruptive than automation misses—which may not be noticed at all, if their consequences are not serious—and this greater salience of automation false alarms than misses may lead the human to perceive the automation as overall less reliable, leading to a general decline in automation dependence with negative implications for reliance as well as compliance (Dixon, Wickens, and McCarley 2007).

Event Base Rate

Finally, the influence of the base rate of to-be-alerted events is noted (Getty et al. 1995; Parasuraman, Hancock, and Olofinbaba 1998). If the need to avoid misses (the cost of misses) remains constant, even when the events to be detected are extremely rare, then a low-threshold setting designed to avoid misses will produce extremely high false alarm rates. That is, the probability that an alarm will sound given that there is no event in the real world will be very high. Using a different statistic, the positive predictive value (PPV) of the alert (i.e., the probability that an alert will actually signal the dangerous event) can be quite low—well under 0.50 (Getty et al. 1995; Krois 1999). Most alerts will be false. As a consequence, it is easy to understand why people frequently turn their alarms off when those alarm systems generate false alarms to events that occur quite rarely (Parasuraman and Riley 1997; Sorkin 1989).

Solutions

The most obvious solution to the issues of human performance costs to imperfect alerting automation is to increase the reliability in discriminating dangerous from safe events. Though this often entails a purely engineering solution (i.e., designing better algorithms), one human factors approach that can sometimes be taken is to reduce the look-ahead time of predictive automation such as that involved in forecasting events or collision warnings. Of course, the look-ahead time should be no shorter than the time necessary for the human to respond appropriately to the alerted state. For example, it makes no sense to have a look-ahead time for a pilot collision warning of five seconds. Though this guarantees high accuracy, five seconds is too little time to maneuver and avoid collision (Thomas and Rantanen 2006).

A second solution is to adopt likelihood alarms, whereby the alert system itself signals its own degree of uncertainty in classifying events as a dangerous signal versus a safe noise (Sorkin and Woods 1985; Xu, Wickens, and Rantanen 2006). However, the operational success of this approach in a dual-task context remains ambiguous (Sorkin, Kantowitz, and Kantowitz 1988; St. John and Mannes 2002; Wickens and Colcombe 2007a,b). A third approach, which is discussed more in chapter 9, is based on the concept of preattentive referencing (Woods 1995). Here the human is given access to continuous information about the evolving state of the alert domain (e.g., the raw data), often in nonfocal sensory channels such as peripheral vision or sonification, which might be used, for example, to represent the continuous sound of a heart beat in an intensive-care monitoring workstation (Watson and Sanderson 2004). Finally, solutions in training can be suggested, typically in training different aspects of attention allocation to be calibrated with actual system reliability as well as training the alarm user to understanding of the inevitable nature of the trade-off between misses and FAs as the threshold is varied and therefore to better tolerate the high FA-rate that must be expected when low base rate events are coupled with imperfect diagnostic automation.

Conclusion

This chapter has discussed properties of events that capture or grab attention, in both the theoretical laboratory-based context and that of the real world, where alarms were the most important application. However, other attention forms of cuing devices, such as highlighting, can be equally relevant to basic research on cuing. In both the theoretical and applied environments, the reliability with which the cue actually directed attention to a meaningful target or event proved to be a critical concept.

We also discussed the characteristics of events and the observer that failed to capture attention, in the context of change blindness. In both successes and failures of attention capture, this chapter's focus was on environmental properties and hence on factors that guide attention in a so-called bottom-up event-driven fashion. Yet we know also that our mind can choose where to attend and often can override these bottom-up effects of salience using top-down or knowledge-driven processes. Thus, the following chapter integrates all the factors responsible for guiding our selection, whether bottom up or top down, and emphasizes the very prominent role of visual scanning in selective attention.

4

Visual Attention Control, Scanning, and Information Sampling

Introduction

The previous chapter demonstrated how attention was controlled by a combination of top-down (i.e., expectancy-driven) and bottom-up (i.e., attention-capture) processes, relying to a considerable degree on relatively basic laboratory research. The current chapter focuses nearly exclusively on visual attention, and because of the heavy—but not total—linkage of visual attention to eye movements, these movements are the subject of discussion here. In real-world environments, how does the brain decide where to sample to obtain and access visual information necessary to perform tasks of interest? In thinking about the travel of selective visual attention around the visual environment, we first describe an analogy of the traveling gaze of attention to our own travel around a geographical environment, before describing eye movement measures and the influences on those movements and then introducing a computational model of visual scanning. We conclude by briefly describing applications to automation complacency and to expertise.

If we consider why a person decides to travel someplace—whether visiting a room, driving around town, taking a longer vacation, or the more virtual form of travel related to visiting a Web site—it is possible to identify a number of influences on the choice of a destination. For example, consider the following:

(1) We may travel places out of habit (e.g., walking straight to the kitchen every morning when you get up, as I do).
(2) We may visit a room next door because we hear a loud noise in the room.
(3) We may visit a location because a lot generally happens there and, hence, we expect to get new information upon our visit—for example, traveling to the local hangout because there is a lot of gossip there or because there are a lot of wise and talkative people who offer good information.
(4) We may travel to a specific destination because we are told there is relevant information there for a specific occasion (e.g., "Go to the library

and check out the latest issue of *Human Factors*, which has an article directly related to your writing.").

(5) Sometimes we may travel to a place simply because of its intrinsic value to us—for example, visiting the mountains to gain positive value. Or we may visit a place to avoid a loss of value—for example, traveling across town to the public transit lost-and-found office to seek a valuable jacket left on the bus.

(6) Often our travel plans are directed by efficiency and economy. If gas is expensive we do not travel far, and we may run several errands in the same neighborhood on the far end of town to avoid several repeated long trips from home—later in the chapter this is referred to as the in-the-neighborhood (ITN) effect. This tendency to avoid long travel will probably be amplified to the extent that the travel is difficult (e.g., our car does not run well; it is rush hour; the vehicle is heavily loaded).

As this list suggests, we may consider any individual traveling someplace (i.e., allocating our body to a location in the world) as guided by the influence of multiple forces combining in ways that we do not fully understand. A corresponding analysis can be made of the forces that influence the allocation of visual selective attention to gain information to update our assessment of a dynamic, evolving situation (Wickens, McCarley, et al. 2007). Here one can parse the influences in a way that corresponds to the aforementioned six factors of physical travel as shown here—influences discussed in detail later in the chapter (Table 4.1).

In applied as well as basic research, it is an article of faith that eye movements are able to provide a reliable index of the allocation of attention. Of course, the correlation between the two variables is not perfect. But regarding the study of information access, across spatially distributed environments such as the airplane cockpit (Fitts, Jones, and Milton 1950), driver environment (Horrey, Wickens, and Consalus 2006; Mourant and Rockwell 1972) or the computer work station (Fleetwood and Byrne 2006) the correlation between eye movements and attention is sufficiently high to as to be valid.

We also note that the movement of the eyes in the control of attention may often be coupled or assisted by other motor activities. For longer travel, (above

Table 4.1   Sources of Influence on Visual Information Access

Source
(1) Habit (procedural scanning)
(2) Attention Capture: Salience
(3) Information Content: Event Rate Or Bandwidth
(4) Information Content: Contextual Relevance
(5) Information Value
(6) Effort Conservation

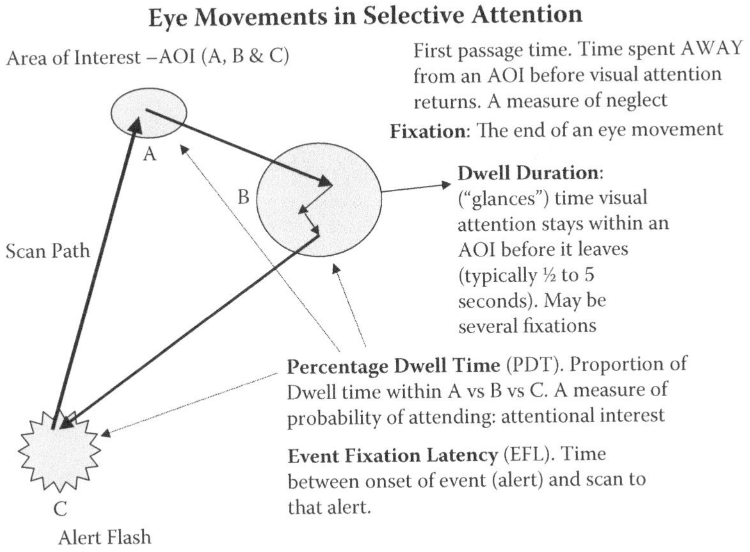

Eye Movements in Selective Attention

Area of Interest –AOI (A, B & C)

First passage time. Time spent AWAY from an AOI before visual attention returns. A measure of neglect

Fixation: The end of an eye movement

A

B

Dwell Duration: ("glances") time visual attention stays within an AOI before it leaves (typically ½ to 5 seconds). May be several fixations

Scan Path

Percentage Dwell Time (PDT). Proportion of Dwell time within A vs B vs C. A measure of probability of attending: attentional interest

Event Fixation Latency (EFL). Time between onset of event (alert) and scan to that alert.

C

Alert Flash

Figure 4.1   Elements of visual scanning.

about 20 degrees) eye movements are assisted by head movements. Sometimes body movements are necessary to bring the eyes to a source of critical information, as when we walk across the room to gain the book in which there is information we must access. With computer interfaces increasingly we use the hands (e.g., controlling a mouse or keyboard) to access screens so that we can attain necessary information. An important concept describe following—related to effort conservation listed in Table 4.1—is the physical effort required to access different sources of information, using whatever mechanism is necessary: eyes, head, body, hands, or even the walking feet.

Eye Movement Measures

As shown in Figure 4.1, measuring visual scanning in operational environments such as the vehicle control or workstation environment, depends on a number of concepts and measures, which are all critical to representing the data of the basic event in scanning, the fixation or dwell shown by the arrowheads in Figure 4.1.

The area of interest (AOI) defined by the analyst is simply an area within which all individual fixations are considered by the analyst to be functionally equivalent. For example, in some circumstances a particular display (e.g., the speedometer or electronic map) might be considered a single AOI; in others the AOI may be a collection of instruments (e.g., the entire dashboard) or another region of space (e.g., the outside world viewed through the windshield).

The dwell duration, sometimes called the glance duration, is the amount of time that the scan remains within a given AOI before departing again—

Transition From:

Transition to:	A	B	C
A	1	2	1
B	2	1	1
C	2	0	1
Total Dwell Time (PDT)	**5**	**3**	**3**

Figure 4.2 Sequential dependencies in attentional transitions of scanning. The figure shows a simple three AOI analysis. The figure illustrates transitions from the AOI shown across the top row to that shown down the side. The figure demonstrates the tight sequential coupling between A and B—tighter than between either of these and C. Bold indicates the total dwell time on each of these across the bottom row. This shows that B and C are equally attended, but this characteristic of the data would not reveal the differences between them in sequential dependencies. Note that the 1s in the negative diagonal represent a transition from an AOI to itself. These auto-transitions correspond to the dwell duration on each AOI.

even though this time may consist of several consecutive fixations on different regions within the AOI, as shown in AOI(B). Since this is typically expressed as a mean value across a number of separate dwells on the same AOI, we refer to it as the mean dwell duration (MDD).

The percent dwell time, or proportion dwell time (PDT), is the total time that the scan remains within an AOI divided by the total time of experimental measurement. Thus, the numerator is the MDD multiplied by the number of dwells on the AOI in question, whereas the denominator is the total time during the work period of interest. As such, the PDT can be thought of as a measure of the relative allocation of attention, or the degree of attentional interest in the AOI in question.

The event fixation latency (EFL) is defined as the latency between the occurrence of a discrete event and the first fixation into an AOI that characterizes this event. For example, it may describe the latency between the onset of a visual warning signal and the first fixation on that signal. However, it may also describe the latency until the first fixation on the instrument in which the out-of-tolerance value was triggered by a nonvisual signal (e.g., an auditory alert). It may be thought of as a measure of noticing time.

The probability of noticing an event is expressed as the ratio of event driven glances to the AOI in question, to the total number of such events. Sequential dependencies, or transition matrices, define the probability that the scan will traverse from one AOI to another. These are often normalized by the absolute probability of the first AOI so that they answer the following question: "Given that the fixation is on AOI1, what is the likelihood that the next fixation will be on AOI2?" These are represented in the 3×3 transition matrix shown in Figure 4.2, which presents what is sometimes called a first-order Markov process. Each entry in a cell represents the frequency of times the eye transitions from the AOI in the column at the top to the AOI in the row along the side. An entry within the negative diagonal represents the frequency with which the eye repeats a consecutive fixation within the same AOI; several such repeated fixations yield a long dwell within that AOI.

The mean first passage time (MFPT) is the average length of time that the eye stays away from a given AOI before it returns (Moray 1986). Thus, it can be used to characterize a period of attentional neglect during which time the human will be vulnerable to missing critical transient events that occur within the AOI in question; consider, for example, how the MFPT for road-way scanning can determine the hazard risk of collisions in driving (Horrey and Wickens 2007).

Limits of Visual Scanning as a Measure of Attention

Of course, visual scanning measures have many limitations in their ability to serve as a tool to index the allocation of limited attentional resources. For example, as noted in the previous chapter, covert attentional movement can often be decoupled from eye movements.

(1) They do not—and by definition cannot—reveal changes in attention allocation within an AOI. This limitation can be addressed to some extent by the analyst reducing the size of the AOI. However, when the size of the AOI is as small as the limited resolution of the eye tracking device to precisely establish fixation position, then the constraint is a hard one. Furthermore, fixation cannot discriminate attention to two different attributes of a single object (e.g., the color versus shape of a colored map symbol; whether the pilot is attending to the bank or pitch of an aircraft while looking at the artificial horizon on the instrument panel).

(2) They cannot easily discriminate the allocation of attention to visual versus nonvisual (i.e., auditory, cognitive) sources unless the eyes are closed. If the eyes are not closed, then they must fixate on some portion of the visual environment even if attention is directed to other sensory modalities or to internal cognitive processes. In particular, the "blank stare" may indicate no comprehension of the information that falls on foveal vision at that moment.

(3) They cannot access the extraction of information from peripheral vision. This issue is particularly important in ground or air vehicle control, given the large amount of relevant information that can be extracted from peripheral motion fields (Horrey, Wickens, and Consalus 2006) or other global characteristics such as the attitude of the true horizon while flying in an aircraft.

(4) Long values of MDD are inherently ambiguous. They could signal the high difficulty of extracting data from a given AOI (e.g., a blurred word or small text), or they could signal extracting a lot of information from

an information-rich source (e.g., a long fixation on an unfamiliar word while reading [Rayner 1998]).

(5) Often the failure to scan an AOI at all is more informative than are differences in positive values of PDT (e.g., differences between 10 and 20 percent). The meaning of a failure to scan is that no information is extracted. In contrast, the meaning of a difference between nonzero values of PDT cannot as easily be interpreted, for it is sometimes possible to extract all of the necessary information from an AOI within one relatively short dwell so that longer dwells or more dwells are unnecessary.

In spite of these limitations, there are many environments in which scanning can provide very informative data regarding selective attention and information seeking. These are environments that are heavily visual, within which AOIs can be relatively well defined and within which the AOIs can be considered to be sources of dynamic information. Such characteristics nicely define the airplane cockpit, ground vehicle control, or workstation environment.

Influences on Visual Information Access

We turn now to a description of each of the six sources of influence on visual information access identified in Table 4.1, describing each in isolation before discussing efforts to model—and, hence, to predict—their influences in consort.

Habit

Certain aspects of visual scanning behavior are controlled by what appears to be habit. Thus, the tendency in most Western cultures to visually search fields from top to bottom and left to right (Wickens and Hollands 2000) appears to relate to reading habits. For a pilot, the standard hub-and-spoke pattern of instrument scanning is taught explicitly in some training programs, is practiced, and, hence, becomes a form of habit of information acquisition (Bellenkes, Wickens, and Kramer 1997).

Salience

As discussed in chapter 3, much basic attention research has addressed the issue of attentional capture. In operational settings, and applied research, the concern is on understanding the characteristics that tend to be more attractive of attention and, hence, that would dominate, override, or outvote the other five influences presented in Table 4.1. These sources of salience— onsets, perceptual salience, and attentional set—were discussed in chapter 3, as was the issue of equal importance: where attention capture fails entirely as reflected in inattentional blindness. Note that event fixation latency is a good way of quantifying attention capture in a nonintrusive manner.

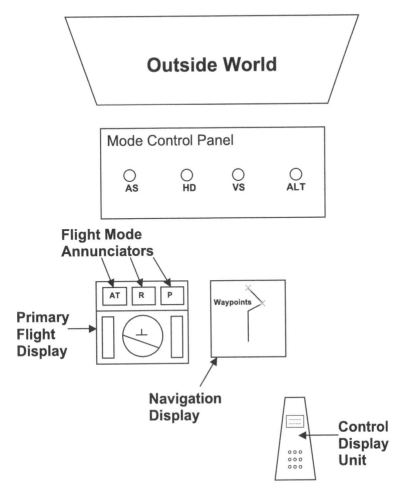

Figure 4.3 The flight management system. The flight mode annunciators, labeled FMA, are small green boxes located on the left panel.

As discussed briefly in chapter 3, a real-world study that illustrates the failures of nonsalient events to capture attention was carried out in the glass cockpit of the Boeing 747-400 simulator, using seventeen highly trained commercial pilots (Sarter, Mumaw, and Wickens 2007). The goal was to understand how these scanning strategies differed from those in the conventional cockpit and, in particular, to gain insight into the symptoms of the mode awareness problem in the glass cockpit whereby pilots are often surprised by—and sometimes are totally unaware of—changes in the mode of operation of the flight management system (Figure 4.3). Changes in the current mode of operation are indicated by the onset of a green box, surrounding the critical flight mode annunciators (FMAs) on the instrument panel. Analysis of the event fixation latency (EFL) to look at an FMA—contingent on a change of state there (signaled by the green box onset)—revealed that the latency was long, measuring several seconds, and on 40 percent of the occasions the scan failed to reach the FMA at all. This finding shows evidence of an attentional neglect of onset events, which might otherwise have been predicted from extrapolating basic research, to show a high success of capture (Yantis 1993).

Information Content: Bandwidth

If a channel, operationally defined here as an AOI, has a high event rate, people will more frequently sample that channel than if the event rate is lower. Indeed, if the amount of information at an AOI can be specified in the language of information theory (i.e., in bits), then the bandwidth of the AOI can be specified as the product:

$$[(\text{bits/event}) \times (\text{\# events/unit time}) = \text{the bandwidth}],$$

typically defined in bits per second (Moray 1986). In practice, if events occur relatively frequently and if those events contain some information, the bandwidth will often be simply expressed as events per second, such as in a driving application, the number of oncoming cars per second, or the number of wind gusts per second (Horrey and Wickens 2007).

In a series of classic studies, Senders (1964, 1980) demonstrated how visual sampling is generally proportional to the bandwidth of the channels being sampled. Thus, if the operator expects a high rate of information at an AOI, she will visit that AOI proportionately more frequently than if she does not. This general finding has been replicated with other augmented tools for information access, such as opening windows on computer-driven displays (Meyer et al. 1999; Wickens and Seidler 1997). Examination of pilot instrument scanning reveals that the attitude directional indicator (ADI), which shows the pitch and roll of the aircraft, is the instrument that changes most frequently given the dynamics of the aircraft (i.e., highest bandwidth) and is also scanned most frequently (i.e., highest PDT) (Bellenkes, Wickens, and Kramer 1997).

Information Content: Context

As described already, it is intuitive that if a channel or AOI frequently presents information because of its high bandwidth, the operator who has learned the channel's statistical properties will expect there to be frequent information and, hence, will sample it frequently. However, a particular AOI, which has a generally low bandwidth, may nevertheless be visited on certain occasions because the context of events that have happened just before suggests that information is now available there. This characterizes the role of attentional cuing discussed in the previous chapter (Posner 1980). Auditory alarms often provide this context. For example, in the operating theater of a hospital the sound of a pulse oxygen alarm will likely direct the anesthesiologist's scan to the pulse-oxygen meter; an engine warning alarm in the cockpit will likely lead visual attention to travel to the engine instruments; or the ping of an e-mail arrival on the computer may divert visual attention to the screen. However, the context may often be driven by information obtained at a prior visual dwell at another AOI. Thus, for example, if a pilot sees the indication of a traffic aircraft on a cockpit traffic display, this should direct visual attention outside the aircraft to scan the sky in the location where the aircraft is suggested to be by the display (Wickens, Helleberg, and Xu 2002; Wickens et al. 2003). Such contextual dependence will lead to sequential dependencies in the statistical analysis of eye movements (Ellis and Stark 1986) (Figure 4.2).

An important aspect of context is the context provided by the momentary qualitative task demands. Thus, if a task requires that an operator compare a commanded system value presented in one display with an actual system value reading in another, it is likely that the eye will travel directly from the former to the latter even if the command indicator does not explicitly suggest or cue that attention should be redirected. Such a contingency will again be reflected in the sequential effects represented in Figure 4.2. This aspect of context that defines task integration will be seen as important as it interacts with information access effort as discussed in the context of the proximity compatibility principle, described later here and again in chapter 7 (Wickens and Andre 1990; Wickens and Carswell 1995).

Both influences (3) and (4) from Table 4.1 then may be generally grouped under the higher level category of expectancy, since both describe the operator's expectancy of obtaining information from a location. Influence (3) characterizes the overall frequency of sampling a single AOI on the basis of its information content (i.e., bandwidth), whereas influence (4) describes the frequency of transitioning between two sources of information, either because one source provides a context that makes the other one relevant or because both sources are temporarily related to the same task.

Information Value

In addition to scanning being driven by the expectancy of obtaining information at a source (influences 3 and 4 of Table 4.1), research also supports—and intuition reinforces—the conclusion that scanning is driven by the amount of information, or value expected to be gained, at the source or AOI in question. Here, *value* can be described in two somewhat different, but related, terms: (1) the formal value of information, in terms of bits (binomial digits in information theory language); and (2) the monetary value, subjective value, or utility that may be derived by knowing information contained at the AOI or, correspondingly, the utility that may be lost by not sampling—a negative utility. We use the latter meaning for value (i.e., utility). Though Senders (1964) did not directly look at utility in his classic scanning experiments, Meyer et al. (1999) did so in an experiment that used the computer window analogy to information sampling, and Sheridan (1970) and Carbonnell, Ward, and Senders (1968) incorporated both concepts of information amount (i.e., bandwidth) and utility in optimum prescription models of how often a supervisor should sample.

To provide an example of the critical distinction between value and bandwidth, for the car driver at night there may be very low bandwidth information in sampling out the window—little is viewable there and with almost no traffic. However, the negative utility of failing to notice the rare roadway obstacle that may appear will, or should, keep the frequency of outside scanning maintained at a high value. Another interesting example of value in information access is the "weapons effect," well known in eyewitness testimony (Steblay 1992; Wright and Davies 2007). Here, when a weapon is used in committing a crime the witness is less likely to remember other features of the perpetrator (e.g., the face), at least in part because attention was directed to the weapon—a valuable thing to notice—and away from other parts of the crime scene.

Importantly, the concepts of information bandwidth (i.e., mapped to expectancy, the probability that information will be observed at an AOI) and of value may be combined in an expected value model of scanning, borrowing from classic models of decision making. (Wickens et al. 2003; Wickens, McCarley, et al. 2007). We may think of the expected value as defining the mental model: characterizing the top-down knowledge-driven factors that both optimally should drive scanning and actually do drive it. Several investigators have examined departures between actual and ideal mental models as reflected in information sampling (e.g., Bellenkes, Wickens, and Kramer 1997; Fisher and Pollatsek 2007; Pollatsek et al. 2006), and some have focused their research attention on the lack of calibration between the true expected value of information sources and the subjective expected value, as the latter is reflected in scan patterns.

Several of these studies in turn have been directly related to roadway scanning, where departures can leave drivers vulnerable to unnoticed hazards. In one pair of such studies, Theeuwes and Hagenzieker (1993) and Theeuwes (1996) asked subjects to search for experimenter-designated target objects within images of traffic scenes displayed on a computer monitor. The investigators found that subjects failed to notice hazards in unexpected locations (e.g., a bicyclist on the inside lane on the left side of the road) and that subjects also failed to scan to such locations.

In another study examining visual scanning on the highway, Summala et al. (1996) noted that a disproportionate number of crashes that occurred between cars and bicycles at T-intersections involved a driver turning right with a cyclist approaching from the right and riding on the left side of the road. This direction—cyclist approaching from the right—is one that would be highly unexpected, as drivers instead would be looking leftward for oncoming traffic before turning. Here again, their research confirmed the failure to look in this rightward direction. Hazards are unexpected there, but failure to notice them can incur a very high negative value.

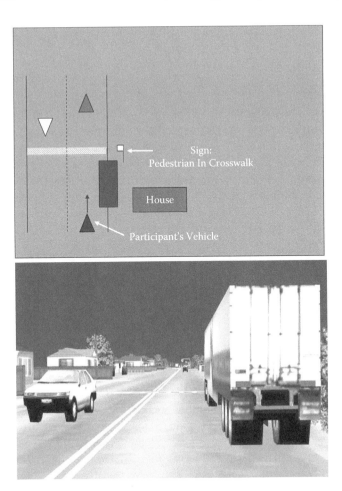

Figure 4.4 Display material used by Fisher and Pollatsek (2007). Top shows a two-dimensional map version used to instruct subjects regarding high valued events (e.g., pedestrian entering the crosswalk). Bottom shows a snapshot of the dynamic three-dimensional scene in which scan training was successfully transferred to improve expectancy for pedestrians.

If people do not look toward low expectancy high value regions in the visual field—thereby departing from expected value prescriptions of scanning—might some form of explicit training be able to help achieve more calibrated scan strategies? Fisher and Pollatsek (2007) and Pollatsek et al. (2006) indeed confirmed that this was possible. They noted several situations where novice drivers failed to scan unexpected but vulnerable parts of their visual field, such as the crosswalk entry hidden behind the parked truck shown in the top of Figure 4.4. Giving subjects' explicit training of the hazard risks of unexpected areas, they found a substantial increase in scanning to these areas when these novice drivers were now immersed within a highly realistic, dynamic driving simulator, encountering similar hazard obscurations in their three-dimensional world, as shown in the bottom of Figure 4.4.

Information Access Effort

Several features of the physical layout of visual space can either help or hurt the redirection of visual attention from one information source to another. These include the visual angle separating the two sources, the extent to which the information lies at different depth planes (requiring reaccommodation), and the amount of clutter (i.e., nonrelevant items) that need to be inspected before the goal of the scan is reached (Wickens 1993). In many circumstances, information must be accessed by other means in addition to eye and head movement. These include the use of manual or vocal interactivity (i.e., keyboarding or voice control in a menu driven display system).

A model of the presumed effort costs of moving attention across different distances as defined by the visual angle is shown in Figure 4.5 (Schons and Wickens 1993; Wickens 1993; Wickens, Dixon, and Seppelt 2002).

At very small angles, no eye movement is required at all—only a shift of internal attention— hence, the cost is low. At angles greater than around two to four degrees, visual scanning must be implemented, but this cost does not increase with larger angles because the eye movement is essentially ballistic: The major cost is simply setting the eye rotation in motion, although there may be a slightly greater effort cost for vertical scans than for lateral ones. At visual angles of source separation beyond perhaps twenty to thirty degrees it becomes necessary to rotate the head, a nonballistic activity in which time and effort cost grows with the amount of movement (Wickens, Dixon, and Seppelt 2002). Finally, this cost begins to increase exponentially, as visual angles are reached where the neck is stretched and body rotation begins to be required. This trichotomization of the visual field was characterized by Sanders and Houtmans (1985) in terms of a no-scan region, an eye field, and a larger head field.

Importantly, within the head field information access is encumbered by a number of other potentially damaging added factors. For example, when engaged in changing three-dimensional motion (aircraft or spacecraft), head movements activate vestibular inputs that produce the disorienting coriolis illusion (Previc and Ercoline 2004). When users wear heavy head gear (e.g., helmets; head-mounted displays), then such head rotation can be particularly effortful and thereby avoided (Seagull and Gopher 1997).

Although simple distance expressed in visual angle is an important contributor to information access effort, several other influences on information access effort can be identified. These include the following:

- The visual accommodation, an important issue when information access must vary between displays at varying distances from the observer

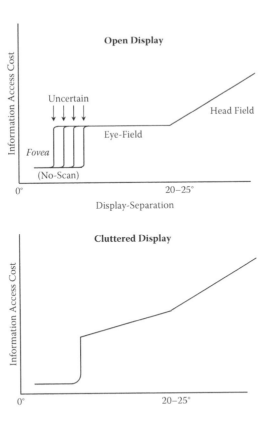

Figure 4.5   A model of information access effort. (Top) an uncluttered open display; (bottom) a cluttered display.

- The effort, particularly time, required to search through information to locate a desired target or destination. As discussed in chapter 5, extensive research has developed fairly nice visual search models predicting that the time required to find a target is proportional to the inspection time per item in the field (I) multiplied by the number of similarly looking nontarget items in the visual field. Note in the bottom of Figure 4.5 that there is an amplified cost to searching through cluttered displays, assuming that the density of clutter is relatively uniformly distributed and that the cluttering items share some common visual features to the destination of the scan.
- The cost of manual interactivity will vary for a lot of reasons. Sometimes this is related simply to computer speed (e.g., How long does it take the computer to access a screen that I request?). Sometimes it is related to the decision time for the user to choose what keys to press (Yeh and Wickens 2001b). With immersed ego-referenced 3-D visual displays, it may be related to the effort required to pan the view (Wickens, Thomas, and Young 2000).

- The cognitive effort required of concurrent nonvisual tasks, the effect of which will be to compete with the effort of information access and to reduce the region of scanning, an issue addressed in detail in chapter 7. For example, Recarte and Nunes (2000) found that the breadth of driver scanning of the roadway ahead is reduced under higher levels of concurrent task workload.

The influence of the effort on scanning and information access has two different manifestations. First, it suggests that people will tend to avoid longer scans or other information access travels when shorter ones can be made. This was described as an ITN heuristic in the context of the travel analogy described at the outset of the chapter. This suggests that when an AOI contains several nested information channels, visual attention will tend to remain there for a while (i.e., a long dwell) before moving on to another AOI. Indeed, research on aircraft cockpit instrument scanning suggests that although pilots look at the general instrument panel, which is one AOI, they tend to sample several instruments within the panel before leaving it to scan outside, which is another AOI (Wickens, Helleberg, and Xu 2002). Second, the effort factor suggests that more effortful scans will be avoided in preference for less effortful ones. This may, for example, lead to fewer scans toward the periphery of a map display (Steltzer and Wickens 2006). The relation between effort and information acquisition is revisited in chapter 7, where we focus directly on the former concept.

It should be noted that the concept of effort in information access interacts with some of the other five factors listed in Table 4.1 in several ways. As one example, it is possible to speak of effort conservation (i.e., moving attention through shorter distances) as gaining utility or subjective value. Doing so allows one to combine the expected value of gaining information from a source (influences C, D, and E) with the anticipated utility costs of that access. This integration has been employed in models of information seeking in decision making (Gigorenze and Todd 1999; Gray and Fu 2004; MacGregor, Fischoff, and Blackburn 1986; Seagull, Xiao, and Plasters 2004; Wickens and Seidler 1997), as described more fully in chapter 7. In particular, in that chapter we describe the combined effects of salience and effort in influencing the *accessibility* of cues used for decision and diagnosis (Kahneman, 2003).

As a second example of this interaction, already alluded to, the contextual influence (D) describes specific task-related forces that impose travel requirements between information sources (e.g., compare a commanded display value with an actual value). Such attention travel links can also be characterized by the effort of the travel. If the cognitive (i.e., working memory) effort required to retain information from the contextual source until the destination is reached is high, then this high load will interact negatively with the distance of travel. This negative interaction will impose greater penalties on

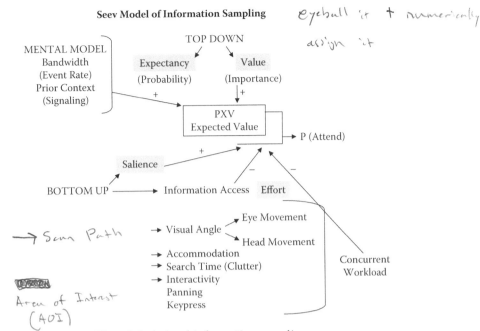

Figure 4.6 The SEEV model of visual information sampling.

the greater separation of information that must be integrated in a task relative to the effort to travel between two sources that are unrelated. Such findings have been repeatedly observed (Schons and Wickens 1993; Seidler and Wickens 1992; Sweller 1994; Vincow and Wickens 1993; Wickens and Seidler 1997) and indeed lie at the core of the display layout principle of functional relatedness (Andre and Wickens 1992; Wickens et al. 1997, 2004). These findings form the most important component of the proximity compatibility principle of display layout and design, discussed more fully in later chapters (Wickens and Carswell 1995). The principle recommends that information sources that must be compared in a task should be positioned close together or in relative locations that require little effort to reallocate attention between them.

The SEEV Trade-Off Model

Thus far this chapter has outlined six influences on where attention and the eye will move, and ample data can be provided to document the influence of each in isolation—and sometimes in combination. However, the applied researcher can benefit from an understanding of how the collective forces of multiple factors will operate together, and it is with this objective in mind that we describe a model called salience, effort, expectancy, value, or SEEV (Figure 4.6), which represents the primary forces that move attention of the skilled operator to selectively attend or sample sources of information (Wickens 2007; Wickens, McCarley, et al. 2007).

Within the model, both probability—bandwidth and cue-driven expectancy—and value can be labeled as top-down or as knowledge-driven forces. As noted already, these may be characterized as embodying the mental model that underlies information acquisition. In contrast, salience and information access effort may be considered bottom up or environmentally driven forces—the former a positive force and the latter indicated by the minus sign, an inhibitory one. To this inhibitory force of access effort is joined the force of concurrent task load (effort, resource demand, as discussed in chapter 7). As shown on the right of the figure, all four forces influence the probability that a source will be attended or that attention will move. The model does not include the separate force of habit because it is assumed that the high level of skill associated with habit will internalize the objective parameters of probability (as reflected by expectancy) and value in a well-calibrated mental model.

A simple form of the scanning model, described as the SEEV model, is one that predicts the probability of attending to an area as follows (Wickens et al. 2003; Wickens, McCarley, et al. 2007):

$$P(A) = sS - efEF + (exEX + vV),$$

where S reflects the strength of salience, EF reflects the inhibitory strength of effort, EX, or expectancy, reflects the collective forces of bandwidth (i.e., event frequency along a channel) and contextual cueing (i.e., event frequency given that an information cue has occurred), and V refers to the value, characterizing the following:

(1) The value of attending to events—or cost of missing them—for their own sake
(2) The value of attending to channels in the context of a task that requires the information in those channels to be integrated with information in other channels
(3) The value, or importance, of tasks served by particular information channels

Within the equation, the lowercase coefficients correspond to the general strengths of these different influences in directing human attention, and the uppercase terms correspond to the levels of these particular parameters in a particular task–display combination.

Three of the four influences of the SEEV model have been validated in five separate scanning experiments: four in aviation (Wickens et al. 2003; Wickens, McCarley, et al. 2007) and one in driving (Horrey, Wickens, and Consalus, 2006). In carrying out such validation we implemented the following six steps:

(1) Define the AOIs, or areas within which fixation can be reliably assessed and which can be unambiguously associated with one or more tasks.
(2) Define tasks served by an AOI and their relative value or importance within the overall multitask context (e.g., in driving, lane keeping could be easily rated as more important than dealing with an in-vehicle navigation system).
(3) Establish the relative relevance of each AOI to each task (e.g., The outside view will be maximally relevant to lane keeping, will have less relevance to the navigational decisions, and will have no relevance to the task of selecting a radio station).
(4) Compute the total relevance of each AOI by summing, across all tasks it serves, the product of task relevance X task value.
(5) Determine the relative bandwidth of each AOI, rank ordered from the most rapidly changing to the slowest.
(6) Define an NXN effort matrix in the form of Figure 4.2, where each cell entry is now the distance between a pair of AOIs; in typical multi-element displays, there are simple heuristics for doing this to establish relative effort (Wickens, McCarley, et al. 2007; Wickens, Sebok et al. 2007).

In all of these operations the term *relative* simply means rank ordering the terms on the lowest integer values possible (e.g., 1, 2, 3).

The model can then produce a dynamic movement of an attentional eyeball to different AOIs, across the workspace, as it is driven by the collective influence of the three forces of effort (negative), expectancy (positive), and value (positive) and from which can be derived the numbers to fill the first-order transition matrix shown in Figure 4.2. Such a matrix can be employed to produce the predicted dwell durations from the cell value on the negative diagonal and PDTs from the row or column totals, respectively. These values in turn can be correlated against actual scanning PDT data, and across the five scanning experiments, cited above, such correlation has been relatively high, ranging from 0.65 to 0.98 (four of these, r > 0.90). Furthermore, because the model generates an actual scan path—or distribution of scan paths—it is possible for the model to provide a measure of the MFPT (attentional neglect) for each AOI (Wickens and Horrey in press; Wickens, Sebok, et al. 2007).

Four additional observations can be made from the collective results of such validation studies. First, in several of these, we have run the model with and without the effort-inhibition term. An improvement of model fit when the effort term is added would suggest that effort reduction is an important component of the actual scanning data. Our findings indicate that such an addition had no influence on the model fit of skilled pilots and, hence, that attention movement of skilled pilots appears to be little inhibited by the greater effort of longer scans. Thus, the scanning behavior of such pilots appears to be optimal in being driven only by the expected value component (Moray 1986).

Second, the degree of fit of scanning model without effort to the observed data accurately predicts the performance of pilots in multitask settings (Wickens, McCarley, et al. 2007). Pilots who are found to be more optimal expected-value scanners performed their tasks better, thus indicating that the model can serve as a sort of gold standard to which scanning can be trained (see chapter 10).

Third, in the driving application (Horrey, Wickens, and Consalus 2006), the model does a much better job of predicting scans to in-vehicle displays than to the roadway in front. This is explainable on the basis of the fact that the in-vehicle displays require foveal vision, whereas a good deal of effective lane keeping can be based on the ambient peripheral visual field (Previc 1998) not well captured by the SEEV model, which captures the direction of foveal vision. These multiple resource properties of ambient peripheral vision are discussed in chapter 8.

Finally, we note that the salience parameter of the SEEV model was not included in these validations. Salience is categorically different from the other three parameters in that it characterizes the unique transient properties of a time-based event rather than the enduring properties of a channel or AOI. This difference implicates the utility of the salience parameter in predicting the event fixation latency (Sarter, Mumaw, and Wickens 2007; Wickens et al. 2005; Wickens, Sebok, et al, 2007), an effort that was not undertaken in the model validation studies previously described.

Applications of the SEEV Model

Here, five important human factors applications of a model of scanning such as SEEV can be summarized.

(1) As applied to training, as we have noted, the scan patterns driven only by expectancy and value can be considered optimal. Hence, scan training can be focused on calibrating expectancy with true bandwidth and value with task importance. In this regard, the training programs implemented by Fisher and Pollatsek (2007) to allow value (V) to expert more influence on novice driver scanning to low-expectancy areas provides such an example.

(2) As one example applied to design, we prescribe that display designers should (a) correlate salience with value so that important events will be likely to capture attention; and (b) correlate distance, or effort, between two AOIs inversely with their joint expectancy so that sources of high bandwidth are close together, requiring minimal attention travel. Furthermore, sources of contextual cues should be close to the attention destination following the cue. Adhering to such a design will allow the parameters of salience and effort to support the optimal mental model (Andre and Wickens 1992).

(3) As a second example applied to design, a figure of merit of a display can be calculated as inversely related to the total scanning distance, as the model attends to task-relevant information on a display layout (Wickens et al. 1997). This will reward display layouts that have frequently used displays close together. Display design tools can be created that continuously calculate this distance as display elements are moved around.

(4) As a safety prediction model, the measure of MFPT or attentional neglect can be calculated for AOIs in which particularly hazardous events may occur. As noted in the previous chapter, when scanning is elsewhere such events, or changes, are likely to go unseen, sometimes even after the scan has returned. Hence, the model can predict the vulnerability to missing critical events, like the roadway hazard, as the scan is directed downward toward an in-vehicle task (Horrey and Wickens 2007; Horrey, Wickens, and Consalus 2006; Wickens and Horrey in press).

(5) A fifth application area of the SEEV model is to the issue of automation complacency (Maltz and Shinar 2003; Metzger and Parasuraman 2005; Moray 2003; Mosier et al. 1998; Parasuraman and Riley 1997). This topic addresses the extent to which operators, dealing with highly reliable automation systems, tend to overdepend on such systems, and in many automation systems, such overdependence has both an attentional manifestation and a related set of consequences (Parasuraman and Riley 1997). Such a phenomenon is opposite to that of undertrust and the cry wolf effect of automated alarms that was discussed in the previous chapter.

The attentional manifestations of this overtrust are that the operator distributes far greater attention to the guidance offered by automation or to other tasks and greatly undersamples the raw data that depict the true behavior of the system that is controlled or monitored by automation (Wickens et al. 2005). For example, the pilot or controller who is overdependent on an automated conflict alert will fail to attend to the actual trajectories of the aircraft in question (Metzger and Parasuraman 2005). The driver who overrelies on a rear-end collision warning system may cease paying much attention to the separation from a vehicle ahead (Stanton and Young 1998). From the standpoint of the prescriptive SEEV model of attention, an irony of such automation modeling—pointed out by Bainbridge (1983) and reinforced by Moray (2003)—is that systems that have never failed in the operator's experience can have a correctly calibrated failure expectancy of zero. Therefore, according to a prescriptive expected-value version of the SEEV model in which expectancy and value are multiplied, this area should never require information sampling—if there are no events there and, hence, zero bandwidth—in spite of the very high value that may accrue from noticing an automation failure. It is for these reasons that (1) we have expressed the SEEV model as the sum of expectancy and value in the equation provided earlier, not the product;

and (2) operators must be trained to expect failures and should even be given false failures during their training so that their monitoring strategies can be calibrated to expect the unexpected (Manzey et al. 2006; Wickens 2001). The importance of attentional scanning measures in the concept of complacency should be noted. This state of complacency will have no behavioral consequences other than scanning until or unless a failure occurs, and at that time the human action may be too late.

Novice–Expert Differences in Scanning

It has long been recognized that major differences between novices and experts in a variety of skills can be characterized by where they look and when. Indeed, since the SEEV model reflects the scanning as driven by the mental model, and since the mental model is a critical component of expertise, the relation between scanning and expertise is self-evident. We will not review the substantial literature in this area (e.g., Lewandowski, Little, and Kalish 2007) but will describe one study in some detail and a second one briefly.

Bellenkes, Wickens, and Kramer (1997) contrasted the two levels of piloting skill—novice versus expert—in a flight scenario involving a series of climbing, turning, and accelerating maneuvers. They confirmed the consistent finding of others that the artificial horizon (ADI) was by far the most frequently fixated instrument on the panel—a characteristic related to at least three factors:

(1) It has the highest bandwidth.
(2) It contains two dynamic attributes, pitch and bank.
(3) It is the primary instrument to support the pilot's most important task of aviating, or preventing stall.

In addition, they found that the MDD of novices on the ADI was nearly twice as long (1 s) as that of experts (600 ms). The novices also fixated on the ADI more frequently than did the experts, so the combined implications of longer MDD and more fixations was that the ADI was a major attention sink for the novices—a very high driver of visual workload. A consequence of this is that novices had relatively little reserve attention to allocate to other instruments. In particular, novices neglected to scan those instruments that appeared to be predictive of the future state (i.e., the vertical speed indicator). They were not as proficient in anticipating aircraft state, and their flight performance suffered accordingly.

Based on these findings of novice–expert differences, three different techniques for training the novices to be more like experts were sought (Bellenkes 1999). In one experiment the eyes of novices were driven to scan more like experts, in essence attracting the eye to move around the instrument panel in a way that would mimic the typical scan pattern of the expert. This passive training of scanning did not prove to be effective. In a second experiment novices were provided with a large amount of part task training in extracting ADI information only, hoping to automatize the extraction of such information so that it would not consume as great an amount of time—shortening dwell duration on the ADI—and thereby would avail more time to sample the other instruments, particularly the vertical speed indicator. Some limited success was produced in this endeavor. In a third experiment novice pilots were simply provided with a more elaborate narrative description of the flight dynamics underlying the changes on the instrument panel. This form of top-down or knowledge-driven training, designed to improve the mental model of flight dynamics, proved to be most successful. Such results reveal that scanning strategies, including those involved in workload management, are knowledge driven, and that providing that knowledge through cognitive understanding of flight dynamics is probably a more secure way to achieve effective scanning than is the imposition of such strategies through bottom-up stimulus-driven training.

Still, scanning training to calibrate the mental model has sometimes shown success (Shapiro and Raymond 1989). In the study described previously, in which novice drivers failed to appropriately sample low-expectancy, high-value areas of the roadway, Pollatsek et al. (2006) found that (1) highly skilled drivers manifest fewer failures; and (2) part task-training simulators, discussed in chapter 10, could be used to train novice drivers to exhibit more optimal scanning in a way that would transfer to a full-task highly realistic driving simulator.

Conclusions

In conclusion, we have described the collective forces that influence eye movements to access information in operational work environments. Often, this collection can be used as a proxy for general movements of selective attention as well, although how to measure such selection within other perceptual modalities (e.g., hearing) remains a challenge. Our discussion here is built on the fundamental building block of attention control, as discussed in chapter 3. It also leads to a critically important application in visual search, which is discussed in the next chapter. Indeed, although eye movements are critical in most search tasks, there are two fundamental differences to justify treating these topics in separate chapters. First, search time and success are measured based on performance, which may sometimes be unlinked from eye movements. Second, our emphasis in the current chapter has been based on acquiring information from specific areas—AOIs—yet visual search, discussed in chapter 5, often involves looking for something, across a field in which the areas of interest can not be well specified.

5

Visual Search

Introduction

Visual search is one of our most common and important attentional skills. It not only pervades everyday behavior but is also a critical component of many specialized tasks. Accordingly, human factors researchers have studied search intensively and across a variety of domains, including driving (e.g., Ho et al. 2001; Mourant and Rockwell 1972), map reading (e.g., Yeh and Wickens 2001b), medical image interpretation (e.g., Kundel and LaFollette 1972), menu search (Fisher et al. 1989), baggage x-ray screening (e.g., McCarley, Kramer, et al. 2004), human–computer interaction (e.g., Fleetwood and Byrne 2006; Fisher and Tan 1989), industrial inspection (e.g., Drury 1990), photo interpretation (e.g., Leachtenauer 1978), airborne rescue (Stager and Angus 1978), and sports (e.g., Williams and Davids 1998). Many basic scientists, moreover, have used search as a window on visual information processing and perceptual representation (e.g., Treisman and Gelade 1980). Thus, researchers have not only garnered an extensive applied knowledge of visual search but also have grounded that knowledge on a strong theoretical foundation.

By definition, visual search is an effort to detect or locate a target item whose presence or position within the search field is not known a priori. Beyond this, however, search tasks differ widely in their characteristics. Two aspects of search are of particular interest. First, search varies in the degree to which the operator knows precisely what to look for. In some cases, no single target object is specified, and the searcher's task is to simply notice objects of potential interest. A driver is expected to detect a detour sign, for example, even if he is not anticipating or searching specifically for it. Similarly, a proofreader is required to search for misspellings without knowing which words will be misspelled. In other cases, the target is a particular item with well-specified properties. This occurs, for instance, when a driver actively searches for a specific exit sign or when the manuscript reader searches for a particular name. The value of target foreknowledge is in enabling top-down or knowledge-driven search processes. When the target is well specified, that is, top-down processes can more effectively guide attention toward likely target stimuli. In the absence of target foreknowledge, search and detection must depend more heavily on bottom-up processes.

Engineering psychologists use the term *search conspicuity* to describe the ease with which an item is noticed when the observer is searching for it and *object conspicuity* or *salience* (a concept discussed in previous chapters) to describe the ease with which it is seen otherwise. Not surprisingly, search conspicuity is typically higher than object conspicuity: we are more likely to notice something if we are looking for it (i.e., an attentional set) than if we are not. Clearly, though, there are many cases in which the operator is not fore-warned to search for a specific and potentially important target (e.g., a detour sign, a pedestrian at night). In creating task-critical signage, labels, or symbol-ogy, the designer should therefore strive to maximize bottom-up conspicuity.

Search tasks can also be characterized by the spatial and temporal struc-ture of the stimulus field. Spatially, search tasks vary from structured to free field (Wickens and Hollands 2000). In the former, the search field is well orga-nized; in the latter, it is open or haphazard. Items in a computer pull-down menu, for example, are rigidly and neatly arranged. An x-ray of a traveler's backpack to be searched by a security screener, in contrast, is likely to be largely disorganized, and a piece of sheet metal to be inspected for defects provides a fully open, unstructured field of search. Search tasks also vary in the degree to which the arrangement of the stimulus field is predictable. The anatomical structures within a chest x-ray, however, for example, may differ in their exact dimensions from patient to patient but will be highly similar in their general form and arrangement. The layout of objects in a baggage x-ray is likely to differ arbitrarily from traveler to traveler. Spatiotemporal structure is useful in that it not only enables more systematic search patterns but can also make the target's position more predictable. Thus, like target foreknowledge, a stimulus field that is structured over space and time can help enable efficient search through top-down processing.

A Model of Applied Search

Researchers from various domains have made efforts at modeling visual search behavior and performance. In addition to providing theoretical insights into the perceptual-cognitive elements of search performance, models of search behavior can be of value to human factors practitioners. In tasks where it is critical to minimize the possibility of missed targets (e.g., searching for defects on a circuit board or potential tumors in a mam-mogram) such a model can provide an estimate the search time necessary to achieve an acceptable level of performance accuracy. In tasks where it is important to minimize search time (e.g., diverting the eyes from the road to scan a dashboard map for route information or to scan a computer menu for options), a model can allow predictions as to whether and when head-down search times will exceed safe limits. Evidence from a number of studies has converged on a general model of applied visual search incorporating a series of perceptual, attentional, and decisional processes (Drury 1975; Kundel, Nodine, and Carmody 1978; Nodine and Kundel 1987). This con-ceptualization has been used to understand and characterize search perfor-mance across a variety of real-world domains and, moreover, posits multiple processing stages analogous to those of more basic theories of visual search

(Itti and Koch 2000; Treisman and Sato 1990; Wolfe 1994). It thus provides a useful framework for understanding applied search and for relating real-world performance to the perceptual and cognitive mechanisms identified by basic research.

The model holds that in conducting a search, the observer first orients to the global stimulus array, assessing the general layout and rough contents of the search field. These processes are carried out simultaneously across the field of search and generally can be completed within a single glance (Kundel and Nodine 1975). They are thus sometimes characterized as preattentive. If the target is highly conspicuous, it will be detected at this stage. Otherwise, the observer will be required to focally scan the image. Although it is possible (Woodman and Luck 2003) and sometimes even beneficial (Boot et al. 2006) to scan covertly, holding gaze in one place while shifting the mental spotlight of attention, scanning will most often be overt, with the searcher executing a series of saccadic eye movements (see chapter 4) to fixate and scrutinize various regions of interest. The goal of scanning is to bring the target within the searcher's visual lobe (Chan and Courtney 1996), the region around the point of regard from within which information is gathered during each fixation. Once this occurs, a successful target-present judgment requires accurate detection, by which the observer matches a pattern extracted from the search field to a stored mental representation of the target stimulus and reaches a yes–no decision by comparing the strength of the match to a response criterion (Green and Swets 1966). A target-absent response occurs when the searcher concludes that the likelihood of finding the target outweighs the cost and effort of further search (Chun and Wolfe 1996; Drury and Chi 1995).

Within this model, a miss (i.e., a false negative response) can occur either because the searcher fails to scan the image adequately and thus does not bring the target within the visual lobe or because the searcher fails to detect the target after bringing it within the visual lobe. Data suggest that the processes of scanning and target recognition are functionally independent. Searchers who perform one of them well, for example, may be poor at the other (Drury and Wang 1986), and task manipulations that improve or degrade one process may leave the other unaffected (McCarley, Kramer, et al. 2004). Efforts to improve search can therefore be targeted at either of the two processes (e.g., Gramopadhye et al. 2002; Wang, Lin, and Drury 1997). Efforts to understand visual search, however, have largely focused on the former process, exploring the nature of the global and focal processes involved in search and in assessing the information that controls visual scanning.

The Parallel–Serial Distinction

As noted, search begins with a global assessment of the search field, followed by a series of attention shifts used to inspect various objects or regions of interest. In the terminology of formal cognitive models, these stages reflect the difference between parallel and serial processing. In parallel search, all items within the stimulus field are processed simultaneously. In serial search, individual items are processed one after the other. One can see here the correspondence of the parallel and serial inspections tasks, with the fully automatic and single-channel behavior, respectively, as discussed in chapter 2. Models of search can also be classified as either self-terminating or exhaustive (Sternberg 1966). Self-terminating models are those in which search ends once a target has been discovered. Exhaustive models, in contrast, are those in which search continues through the full search field even if a target is discovered early. Baggage x-ray screening provides a real-world example of self-terminating search; as soon as any threat object is detected in a passenger's bag, visual search ends and the passenger is pulled aside. Conversely, mammography provides an example of exhaustive search. Since the course of treatment depends on knowing how many lesions are present, search of the mammogram has to continue even after a single lesion is discovered.

The standard form of the parallel model assumes that the time needed to find a target item in the search field is *not* affected by the set size, the number of items (nontargets + targets) contained within the field. In other words, the mean processing time per item remains the same regardless of how many items are in the search field, a characteristic known as unlimited capacity. The standard form of the serial model assumes that the average processing time is the same for all items within the search field (Townsend and Wenger 2004). Search on target-present trials can be either self-terminating or exhaustive, regardless of whether processing is parallel or serial, but both the standard parallel and the standard serial model assume that exhaustive processing is required to determine that a target is absent—that is, all items must be inspected. Given these assumptions, the conventional method of attempting to distinguish between serial versus parallel and self-terminating versus exhaustive models is to examine the effects of set size on the response time (RT) for yes (target-present) and no (target-absent) responses (Sternberg 1966; Townsend and Ashby 1983). These relations are shown in Figure 5.1. Set size is manipulated by keeping the number of targets in each display at zero (target absent) or one (target present) while varying the number of nontargets—different predictions arise when more than one target is present.

When the search is self-terminating, the standard parallel model predicts that target-present RTs will be independent of the number of nontargets within the search field. In other words, the RT × set-size function will be flat (Figure 5.1, bottom left, solid line). This flat slope is a key diagnostic used for testing theories of unlimited capacity parallel processing and is also, of course, an applied goal in the design of any display or environment in which visual search is carried out. When a target is absent or when search is otherwise exhaustive, (e.g., target is absent) the standard parallel model predicts that RTs will increase as a concave function of set size (Figure 5.1, bottom left, dashed line). The reason for this increase is that RT for exhaustive search

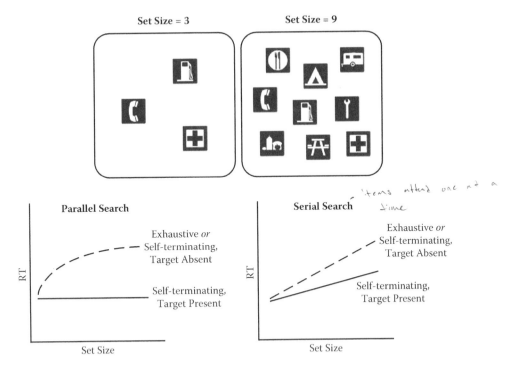

Figure 5.1 The technique for distinguishing serial from parallel search is to vary set size and to measure the effects on RT. The top panel illustrates displays from two set-size conditions in a visual search task using traffic icons as stimuli. The observer's task is to determine whether a particular icon (e.g., a first-aid symbol) is present among varying numbers of nontarget icons. The bottom panels represent RT patterns from different forms of search model. The standard parallel model (bottom left) predicts that RTs for self-terminating target-present search (solid line) will be unaffected by set size—a flat slope—whereas RTs for exhaustive search or for self-terminating target-absent search will increase as a concave downward function of set size (dashed line). The standard serial model (bottom right) predicts that RT will increase linearly with set size. If search is self-terminating, the slope of the search function will be twice as large for target-absent trials (dashed line) as for target-present trials (solid line). If search is exhaustive, target-present and target-absent trials have the same slope.

is determined by the processing time for the slowest item within the search field. Because of random variation in the processing time for individual items, the likelihood of one or more very slow finishing times increases as items are added to the field. Thus, even in the standard parallel model, RTs can increase for purely statistical reasons.

The standard serial model (Figure 5.1, right side of graph) predicts a linear increase in RT as a function of set size, since a larger number of items will have to be inspected each trial. The self-terminating serial model predicts that the slope of the RT × set-size function will be twice as large for target-absent as for target-present trials, since when a target is present the searcher will, on average, need to inspect only half of the items within the display before finding it, whereas when the target is absent, all items will always need to be searched. The exhaustive serial model, conversely, predicts that RT slopes will be equivalent for target-absent and target-present trials, since the number of items (N) inspected will be the same in either case. Finally, when the positions in a display are consistently searched in the same order—as, for example, when a searcher scans columns from top to bottom when looking for a name in the phone book—the serial self-terminating model also predicts serial order effects, with detection times being proportionately shorter for targets in early positions than for those in later positions (Neisser 1963).

The properties of the serial self-terminating search model just described allow one to approximate search time as follows:

(1) $RT = a1 + bN/2$ when target is present and
(2) $RT = a2 + bN$ when target is absent,

where b is the time to inspect each nontarget item, a1 is the sum of the time to decide that the target is identified and execute a response, and a2 may represent the time required to assure that a target it not present. Although these equations or computational models do a reasonably good job accounting for search times in some realistic environments (e.g., Nunes, Wickens, Yin 2006; Remington et al. 2001), their accuracy remains far from perfect. As an important aside, it should be noted that auditory phone menus must always be searched in a serial fashion, and diminishing their efficiency further is the fact that the constant b is always going to be long—the time required for the phone menu to speak each option.

In practice, the problem of characterizing visual search is often more difficult than the simple taxonomy of the standard parallel–serial and self-terminating–exhaustive processes suggests. The straightforward predictions just described mask complications that make experimental efforts to distinguish different classes of models challenging. The prediction of flat RT functions from parallel search, for example, is based on the assumption of unlimited processing capacity—that is, the assumption that mean processing rate per item is unaffected by the number of items in search field. If processing is limited in capacity such that processing rates slow down when set size increases, then adding nontargets to a display will increase yes RTs even if search is parallel and self-terminating. Under certain conditions, in fact, a limited capacity parallel model can perfectly mimic the predictions of a serial model (Townsend and Ashby 1983).

The dichotomous classification of search as self-terminating versus exhaustive search is also oversimplified, ignoring the possibility that an observer might terminate search early with an informed guess or might search beyond exhaustively by scanning some items more than once, a strategy that is reasonable when target detection is difficult or the costs of a missed target are high (Chun and Wolfe 1996; Horowitz and Wolfe 1998). Finally, empirical discoveries and theoretical advances have made it clear that search tasks do not always fall neatly into dichotomous parallel versus serial categories. Real search behavior, as the aforementioned applied model suggests, appears to reflect a complex of interwoven, interacting parallel and serial processes. Nonetheless, the parallel–serial dichotomy has helped to delineate components of more complex search behaviors, has provided the context for much of the work that cognitive psychologists have done on visual search, and provides a useful point of entry to an understanding of current search theory.

Parallel and Serial Processes in Search

Evidence for distinguishable parallel and serial processes in visual search can be traced back to very early work in cognitive psychology. Green and Anderson (1956) measured RTs as subjects searched through matrices of colored two-digit numbers looking for a designated target number. When subjects were not told what color the target would be, RTs were determined by the total items within the display, consistent with a serial search. When subjects were informed of the target's color, however, the number of non-targets of different colors had very little effect on RT. In other words, when searching for a target of a specific color, subjects seemed to filter away items of different colors in parallel, a finding that presaged modern models of guided visual search to be discussed later in the chapter (e.g., Wolfe 1994). Neisser (1963) asked subjects to search columns of text for a line that contained a designated target character. RTs were analyzed as a function of the target line's position within the list. In general, RTs showed a serial-order effect suggesting a serial self-terminating, top-to-bottom search strategy. The slope of the RT × position function, though, was steeper when the target letter and distractors were similar in appearance (e.g., a Z target among angular distractors, like K and X) than when they were dissimilar (e.g., a Q target among angular distractors K and Z). Neisser (1967) proposed that coarse differences in target and distractor features can be detected in parallel but that finer discriminations as between Z and X require or serial processing with focused attention.

Work by Anne Treisman and colleagues elaborated on Neisser's (1963, 1967) findings and couched them within an elaborate theory of visual perception and attention. Treisman's feature integration theory (FIT) (Treisman and Gelade 1980) proposed that complex objects are represented in the visual system as conjunctions of rudimentary properties. Simple attributes—color, motion, orientation, coarse aspects of shape—are encoded preattentively in mental representations known as *feature maps*, where a single map is selective for a particular value (e.g., red) along a given stimulus dimension (e.g., color). A multidimensional object like a large red square is represented by the pattern of activity it evokes across the array of feature maps. Representation is constrained, however, by the need to coordinate this activity. In the initial stages of processing, multiple feature maps are not spatially registered with one another. Focal attention is therefore necessary to appropriately bind together the various features of an object. Only by focusing attention on a single location can the observer link activity at that position across multiple maps, producing an accurate representation of a multifeature object. Outside of focused attention, features from multiple objects can be jumbled or combined inappropriately. Quickly viewing a display that contains a red square and a blue cross, for example, the subject may perceive a red cross and blue square. Such inappropriate combinations of features are termed illusory conjunctions (Treisman and Schmidt 1982).

FIT thus proposes that perceptual processing of visual features (e.g., the color red) is qualitatively different from processing of feature conjunctions (e.g., a red square). To detect the presence of a single feature target, it is sufficient simply to check for activity in the relevant feature map. This allows rapid parallel search with no need for focused attention—a flat slope. To detect a specific conjunction of features, it is necessary to scan the display one item a time, binding together stimulus properties.

To illustrate the differences between feature and conjunction processing in an applied context, consider the design of an electronic traffic map like that used by some on-line newspapers. Different kinds of events are represented by different shapes, with circles denoting traffic build-ups and diamonds denoting accidents. The severity of each event is represented by symbol color, with green indicating light, orange indicating moderate, and red indicating severe. Under this system of coding, according to FIT, it is possible either to detect events of a certain kind by preattentively registering symbols of a particular shape or to detect events of a given severity by preattentively registering symbols of a particular color. However, to detect an event of a particular kind and severity (e.g., a moderately severe accident) requires focal attentional scanning to link the colors and shapes of individual symbols. Hence, the map needs to be scanned serially, item by item, to find, say, a build-up of modest severity. An important implication of the feature–conjunction distinction is that where possible, icons, signs, alerts, and symbology that require rapid detection and recognition should be distinguishable on the basis of a single visual feature. Data have confirmed that search times for computer icons increase as icon complexity, as defined by the number of stimulus features, increases (McDougall, Tyrer, and Folkard 2006).

Attentional Guidance

Treisman and Gelade's (1980) report introducing FIT quickly became a landmark in cognitive science. Research further testing the predictions and elaborating on the details of the theory followed (see Quinlan 2003 for a comprehensive review), and a number of additional models incorporating the central premise of FIT—that a scene is initially encoded as a set of primitive perceptual features—emerged (e.g., Itti and Koch 2000; Wolfe, Cave, and Franzel 1989). The most important qualification to Treisman's original idea produced by this work was the demonstration that search is systematically guided by scene properties and the observer's task set. Echoing the findings of Green and Anderson (1956), Egeth, Virzi, and Garbart (1984) found that observers performing a conjunction search can restrict attention to stimuli that possess at least one of the target properties. In searching for a red O among black Os and red Ns, for example, observers in Egeth, Virzi, and Garbart's study limited their search to the red items, effectively filtering away all black distractors in parallel. This phenomenon can be described as *guided search*: search that is restricted to items similar to the target. Wolfe, Cave, and Franzel (1989) revealed further evidence for search guidance by demonstrating that search for a conjunction of three features (e.g., color × size × shape) is faster than search for a conjunction of only two features (e.g., color × size, color × shape, or shape × size), an effect suggesting that search becomes more efficient when attentional set can be tuned more finely on the basis of additional feature information.

To capture the phenomenon of attentional guidance, computational models have incorporated the concept of visual salience, a notion introduced in the previous chapter. As described by Koch and Ullman (1985, p. 221), salience is a "global measure of conspicuity" at any given point within the visual field, as determined by the confluence of visual features at that position. The distribution of salience across the visual field thus provides a "'biased' view of the visual environment, emphasizing interesting or conspicuous locations" (ibid.). During search, attention is attracted to points of high salience. A target that generates a large salience peak relative to the background will therefore be detected easily. A target that is not highly salient relative to the other items in the search field will demand focal scanning, with search beginning at the items of high salience and gradually working toward items of lower salience. As discussed in chapter 3, salience is largely determined bottom up by contrast between feature values (e.g., the contrast between different colors on either side of a chromatic border). Figure 5.2 illustrates the distribution of bottom-up salience across a visual scene, as calculated by the salience model of Itti and Koch (2000).

Though bottom-up salience will bias search, knowledge-driven processes can also amplify or attenuate the activation within feature maps, allowing the searcher to bias attentional scanning toward those objects within a scene that contain known target features (Treisman and Sato 1990; Wolfe 1994). This top-down influence explains the findings of search selectivity reported by Green and Anderson (1956), Egeth, Virzi, and Garbart (1984), Wolfe, Cave, and Franzel (1989), and others. Comparisons indicate that a computational model guided solely by bottom-up salience calculations can do a reasonable job of simulating human search behavior (Itti and Koch 2000) but that

Figure 5.2   The distribution of bottom-up salience in a traffic scene as estimated by the model of Itti and Koch (2000). Dark spots on the right panel indicate points of high salience within the scene depicted on the left. Points of high salience correspond to points of high visual contrast. In the depicted scene, salience peaks occur at the location of the foreground traffic signs and, to a lesser extent, the treetops protruding above the horizon on the left—objects that are high in luminance contrast with their background. If visual search were guided strictly by bottom-up salience, these would be the first objects inspected.

models incorporating a top-down component perform better (Navalpakkam and Itti 2005).

What Determines Search Efficiency?

Note that a model integrating a salience map with a mechanism of focal scanning is neither strictly parallel nor strictly serial; shifts of the attentional focus occur in series, but calculations of salience are performed in parallel across the visual field. Consequently, search tasks vary along a continuum of ease, and it is generally not possible to classify a given task as being either strictly parallel or strictly serial. Modern convention is therefore to describe search as being more or less efficient or inefficient, eschewing talk of discrete parallel versus serial categories (Wolfe 1998). Wolfe and Horowitz (2004) summarize a number of findings on the mechanisms that determine search efficiency. One fundamental conclusion of this literature is that search for a target defined by a single basic visual feature is often highly efficient, approaching the flat-slope performance expected of a parallel, unlimited capacity processor. Properties that appear to constitute basic visual features include color, motion, orientation, and size and spatial frequency (Wolfe 1998). Thus, for example, a red target pops out from among green distractors, and a moving target is found easily among static distractors. Treisman and colleagues (Treisman 1986; Treisman and Gelade 1980; Treisman and Souther 1985) have discussed various techniques for determining what visual properties do and do not constitute basic features.

Visual search for feature targets is not uniformly efficient, however, nor is search for conjunction stimuli invariably difficult. Search efficiency is modulated, for example, by the similarity between targets and distractors (Duncan and Humphreys 1989; Wolfe and Horowitz 2004), such that even search for a basic visual feature can be slow if the target and distractors are highly similar. Moreover, the target-distractor differences necessary to produce efficient search are larger than the differences needed to discriminate between stimuli when they are in the focus of attention (Nagy and Sanchez 1990). Thus, a red target that pops out from among green distractors may be found only with difficulty among orange distractors, even if the red and orange stimuli are easily distinguishable in attentive vision. Dissimilarity among distractors can likewise impair search efficiency, with search generally becoming more difficult as the distractor items become more heterogeneous (Duncan and Humphreys 1989). Thus, as shown in Figure 5.3, search for a text-file icon on a computer desktop screen is likely to be faster if the other icons cluttering the screen are all of one type, such as PDFs (Figure 5.3, top) than if they are of multiple different types (Figure 5.3, bottom).

Data also indicate that even when target-distractor similarity is held constant, it is easier to detect an item that possesses a unique feature than it is to detect an item that lacks the feature among distractors that possess it (Treisman and Gormican 1988; Treisman and Souther 1985). That is, the presence of a property is more conspicuous than the absence of it. For example, the letter Q is distinguished from the letter O by the presence of a tilted line segment. It is therefore easier to search for a Q among Os (Figure 5.4a, left) than to search for an O among Qs (Figure 5.4a, right). Such asymmetries can also arise when targets or distractors are distinguished by a deviation from a minimum or default value along some feature dimension. More specifically, search is easier when the target deviates from the minimum value than when the distractors do. Vertical and horizontal appear to be coded by the visual system as default values of line orientation, for example (Foster and Ward 1991). Thus, search for a tilted target among vertical distractors is generally efficient (Figure 5.4b, left), whereas search for a vertical target among tilted distractors is slow (Figure 5.4b, right). Novelty, or unfamiliarity, appears to act as a salient property such that it is easier to search for an unfamiliar target among familiar distractors than to do the opposite (Wang and Cavanagh 1994). For example, a reversed letter is detected easily among correctly oriented distractors (Figure 5.4c, left), but a target letter in its correct orientation is found only with difficulty among reversed distractors (Figure 5.4c, right). The exact cause or causes of these and other search asymmetries remains a topic of debate (Rauschenberger and Yantis 2006; Wolfe 2001), but the effects themselves imply that display symbology should be designed such that high-priority targets are coded by the presence of a unique or novel property. The importance of this principle is illustrated by an aviation accident that occurred when a pilot failed to realize that the air-

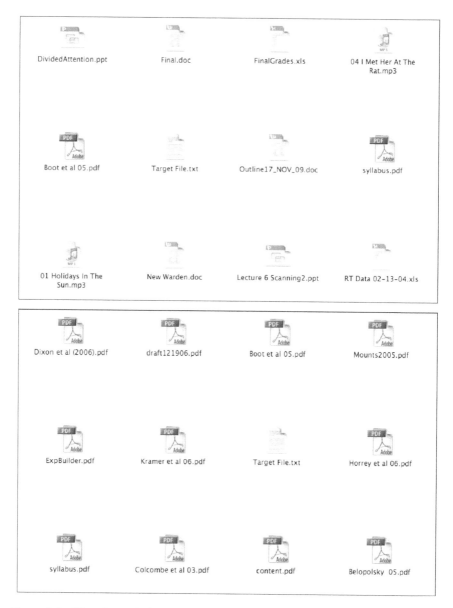

Figure 5.3 Visual search for a target object (the target file) is harder when distractors are heterogeneous (top) than when they are homogeneous (bottom).

port he was approaching was not equipped with radar. The only indication that the airport was not radar equipped was the absence of a letter R from a cockpit chart; by convention, radar-equipped airports were tagged with an R. The data on search asymmetries suggest that a more effective display strategy would be to code important safety critical information such as

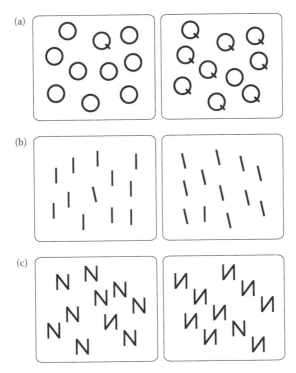

Figure 5.4 Examples of search asymmetries. In all three panels, it is easier to detect the unique item in the left panel than in the right panel despite the fact that the similarity between targets and distractors is the same in both cases.

the absence of radar with the presence of a unique symbol or feature (e.g., Fowler 1980).

Finally, the search for several possible targets is harder, or longer, than the search for a single target. This is because each item inspected must now be compared against a mental representation of two or more desired targets, and this memory search is also often a serial time-consuming process (Schneider and Shiffrin 1977; Sternberg 1966).

The following summarizes the several effects discussed on target search speed. Search is easier, or faster, if the following are true:

(1) The target is uniquely defined by one level on only one sensory continuum.
(2) The level defining the target is not similar to those of the nontargets, or distractors.
(3) The distractors are homogenous.
(4) The target possesses a unique feature rather than uniquely lacking the feature.
(5) The target—not the distractor—differs from a default value, such as verticality.

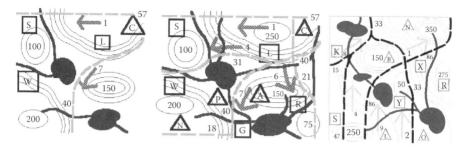

Figure 5.5   Examples of low-clutter (left) and high-clutter (center) maps used by Yeh and Wickens (2001b). Search times increased as an approximately linear function of the number of markings on the map, which is less on the left than in the center with a one-second cost for the more cluttered map. In the right panel, lowlighting was used on the geometric symbols and the arrows to allow search through the higher-contrast water features (i.e., lakes and rivers) to proceed with no cost of the other elements.

(6) The target—not the distractor—is novel.
(7) Only one target needs to be sought at a time.

Attention Guidance and Display Design

Various authors have employed the theoretical findings previously described to inform the study and design of maps and other complex displays, seeking techniques to guide attention to a subset of items assumed to be most relevant to the user's task. In real-world scenes and displays, clutter slows visual search in the same way as added distractors in a conventional laboratory search task (e.g., Ho et al. 2001; Nunes, Wickens, and Yin 2006; Remington et al. 2001; Teichner and Mocharnuk 1979). The time needed for a controller to detect an air-traffic conflict, for instance, increases as an approximately linear function of the number of aircraft within the display (Remington et al. 2001). Similarly, the time necessary to locate an item on a digital map increases as a near-linear function of the number of onscreen markings (Yeh and Wickens 2001b) (Figure 5.5). The display designer's goal is to enable efficient search despite potentially high levels of clutter. One way to do this is to encourage attentional filtering of clutter based on visual features. Consistent with findings from the basic search literature (Egeth et al. 1984), color coding appears to be especially effective for this purpose. Remington et al. (2001) found that detection time for a traffic conflict in an air-traffic control display is greatly reduced when color is used redundantly with text to code altitude, allowing controllers to scan for conflicts only between aircraft in the same altitude range. Yeh and Wickens (2001b) found similarly that color coding of different object classes on a digital map like that shown in Figure 5.5 allowed for about one second faster acquisition of target objects.

An alternative to color coding, suitable for use in monochromatic displays and with system operators whose color vision is deficient (7 percent of males), is intensity coding, the rendering of information at different levels of luminance contrast. This can involve highlighting of high-priority information (sometimes called backgrounding) of low-priority information (Fisher et al. 1989; Kroft and Wickens 2003; Wickens, Alexander, et al. 2004). This is illustrated in the right panel of Figure 5.5, in which the geometric symbols and arrows have been lowlighted compared with their counterparts in the other two panels, thereby allowing easy focus of attention on the remaining highlighted items—roads, lakes, and rivers. In the most extreme case, low-priority information may be removed from the display entirely, a manipulation known as decluttering (Kroft and Wickens 2003).

Although intensity coding and lowlighting may be less effective than color coding for operators with normal color perception (Yeh and Wickens 2001b), it nonetheless can aid visual search. Using maps such analogous to those shown in Figure 5.5, Wickens, Alexander, et al. (2004) examined the effects of information highlighting or lowlighting on visual search in a simulated vehicle dispatching task. Observers searched through monochromatic city street maps for vehicles and destinations marked by alphanumeric characters. Their task was to answer questions that required them to locate various vehicles and destinations on the map. The relative luminance of the vehicles and the map—streets, buildings, and destination markers—varied across conditions. Data showed that performance was best when the vehicle and destinations were presented at different luminance levels, suggesting that intensity coding can facilitate search in much the same way as color coding.

Lee and MacGregor (1985), Wickens, Alexander, et al. (2004), and Fisher et al. (1989) developed computational models that build on the serial search model represented in Figure 5.1 to predict the benefits of highlighting or lowlighting on map search and computer-menu search, respectively. In particular, lowlighting can effectively eliminate the unwanted contribution of those noise elements if it is known that the lowlit elements do not contain the target. If instead these elements are unlikely to contain the target but may occasionally do so, then issues of cuing validity become relevant. These circumstances occur when an intelligent agent (e.g., automation) makes an inference that all targets the searcher wishes to find share certain features and, hence, the agent will highlight all items showing those features. But in the natural world, it is possible that a desired element does not contain these features (i.e., the automation is wrong). These issues of cuing reliability are discussed in chapter 3 (Fisher et al. 1989).

Spatial Grouping and Search

Not surprisingly, data indicate that search efficiency is mediated by scene organization such that search is easier within a well-structured arrangement of items. Spatial structure can aid search in at least three ways. First, grouping of related items minimizes spatial uncertainty. After discovering the relevant group of items, the searcher can restrict attention to that region of the display and avoid the need for further scanning between groups. Second, spatial proximity can allow and encourage the searcher to process multiple items within each group in parallel (Treisman 1982). Finally, a well-organized structure may allow better tracking of what items have and have not been searched (McCarley, Wang, et al. 2003; Peterson et al. 2001), a particular advantage when the search task may be susceptible to repeated interruptions. Niemela and Saarinen (2000) illustrated the benefits of grouping to visual search in a task using computer desktop icons similar to those shown in Figure 5.3 as stimuli. People were asked to search for an icon of a specified type (e.g., a Microsoft Word document) with a designated name. Search times were significantly faster when icons were grouped by type than when icons of different types were randomly interspersed. Note that in this study, since icons of different types differed in their appearance, visual features were also available to help to guide attention toward the appropriate group of objects. In other instances, such as when the items to be searched are text strings grouped on the basis of semantic properties, there may be no distinctive differences in appearance between sets of items. In these cases, easily discriminable labels attached to the groups will allow the searcher to find the relevant set of items quickly (Hornof 2001).

Contextual Constraints on Search

Spatial grouping creates structure within a given scene or display. In the natural world, structure also arises from similarities that exist over time and across different scenes. Objects within most real-world scenes are not randomly distributed but are constrained to appear at somewhat predictable locations within predictable settings. Traffic-control signs typically appear on the driver's side of the road; a set of car keys is more likely to be found on the countertop than on the floor. As might be expected, such spatial constraints play an important role in guiding attention during visual search. Very simply, an item will be found more easily in a contextually appropriate setting and location than when it appears out of its proper context (Biederman, Glass, and Stacy 1973). Knowledge of contextual constraints—where they exist—is thus an important component of visual search expertise. Hoyer and Ingolfsdottir (2003), for instance, compared performance of novice and expert medical technicians in visual search for bacteria morphology in Gram's stain photomicrographs. On each trial, the subject was shown a bacterial specimen as a target and then was given a stain in which to search. The authors manipulated the relationship between the target specimen and the background stain. On some trials, the target specimen was shown within a stain of the appropriate diagnostic category. On other trials, the target was presented within a stain of a different category. Novice technicians performed equally well whether or not the target specimen and the search field were appropriately matched. Experienced medical technicians, however, found the target more rapidly when the target and search field were congruent than when they were incongruent. Targets were easier for the experts to find and recognize, that is, when they appeared within the context with which they were normally associated.

Research by Chun and colleagues provided insight into how contextual constraints affect search. Experiments were cleverly designed to examine the influence of predictable stimulus properties independent of semantic knowledge. Stimuli were haphazard arrangements of simple characters. In one experiment, for example (Chun and Jiang 1998, experiment 1), subjects were asked to find a T-shaped target among L-shaped distractors, a difficult serial search task that demands careful attentional scanning. The spatial arrangement of items within each display was generated randomly, but a subset of the stimulus arrangements appeared repeatedly throughout the course of the experiment, interspersed with novel stimulus layouts. Within a repeated layout, the target's location was always the same. Subjects quickly learned to exploit this spatial predictability; after only a small number of exposures, RTs for repeated layouts were shorter than those for novel layouts. Later experiments (Chun and Jiang 1999) found similar benefits driven by consistent pairings of target and distractor identity across trials and by the predictability of target and distractor paths within displays of moving stimuli. Chun and Jiang (1998) described these effects as contextual cuing.

Interestingly, postexperimental tests found that subjects were not consciously aware of the contextual information that influenced their search behavior in Chun and Jiang's (1998, 1999) experiments. They were no better than chance, for example, in discriminating repeated from novel layouts in the spatial context experiments. The attention-guiding benefits of context in these studies thus appeared to reflect the operation of an unconscious, or implicit, memory system. Chun and Nakayama (2000) suggested that the value of implicit attentional guidance is in allowing efficient visual search while minimizing demands on conscious cognitive processing. The implicit, nonsemantic nature of contextual cuing suggests that learners attempting to develop full expertise in a search task may sometimes be unable to do so through verbal instruction alone. Even medical students with many hours of classroom training in radiology, for example, fail to show the strategic scanning behavior evident in expert radiologists' eye movements when searching for abnormalities in a chest radiograph (Kundel and LaFollette 1972). Thus, implicit contextual cuing appears to reach its full strength only through repeated exposure to stimulus exemplars. After developing their skills, conversely, experts in a search task may be unable to fully verbalize their strategies or to describe the information that guides their attentional scanning (Ericsson and Simon 1993). It is unlikely that knowledge of the visual properties of various diagnostic categories in the Gram's stain search task studied by Hoyer and Ingolfsdottir (2003), for example, could be acquired through explicit instruction alone or verbally expressed in great detail.

Of course, it is not the case that top-down or memory-based attentional guidance is entirely implicit. Explicit knowledge or instruction can direct scanning as well. For example, as discussed in the previous chapter, the study by Pollatsek et al. (2006) revealed that novice drivers could be trained to demonstrate safer roadway scanning if their attention was directed to high-risk areas that they had otherwise overlooked.

The Visual Lobe and the Useful Field of View

Unless a target object is conspicuous enough to be detected with global parallel processes, search will require serial scanning with focal attention. Exactly how focused, though, is focal attention? In other words, how big is the visual lobe? As noted already, the visual lobe—sometimes called the useful field of view (e.g., Ball et al. 1988), functional field of view (Pringle et al. 2001), or eye field (Sanders 1963)—is the area surrounding the point of regard from within which information is processed each fixation (it is not to be confused with the angle of foveal vision and is often larger than the four-degree angle usually associated with the fovea). The visual lobe is typically measured by presenting a target object at different distances from the point of fixation, holding the eyes fixed, and determining the distance at which some measure of visual performance (either target detection, identification, or localization) falls below a criterion level.

As might be expected, the exact size of the visual lobe varies with stimulus characteristics and task demands. In general, the visual lobe is smaller when the target is smaller, inconspicuous, or appears embedded among distractors than when it is large, salient or appears by itself (e.g., Jacobs 1986; Mackworth 1965). The visual lobe also tends be smaller in older than in younger adults (e.g., Scialfa, Kline, and Lyman 1987; Sekuler and Ball 1986) and shrinks when the observer is placed under high levels of stress (e.g., Bursill 1958; Easterbrook 1959; Weltman, Smith, and Edstrom 1971) or cognitive load (Atchley and Dressel 2004; Ikeda and Takeuchi 1975; Williams 1982). Note that the finding that a nonvisual secondary task (e.g., an auditory secondary task) can shrink the visual lobe implies that lobe size is at least in part a measure of attentional breadth and not simply a measure of sensory function (Atchley and Dressel 2004). Consistent with this speculation, lobe size tends to be poorly correlated with variations in foveal visual acuity across people (Chan and Courtney 1996).

The size of the visual lobe affects search performance because it determines how carefully the observer must scrutinize the search field. A large visual lobe enables the observer to process a greater portion of the image with each gaze, ensuring that fewer eye movements will be required to blanket the field (Kraiss and Knäeuper 1982). Accordingly, visual lobe size is correlated with search efficiency. Among photo interpreters, for example, visual lobe size strongly predicts search speed and detection rate for targets hidden in aerial imagery (Leachtenauer 1978). Likewise, searchers with a large visual lobe tend to show shorter target detection times in industrial inspection tasks (Gramopadhye et al. 2002), are quicker to notice events within cluttered real-world scenes (Pringle et al. 2001) and, at least among older adults, appear be at reduced risk of being involved in a traffic accident (Owsley et al. 1998).

The relationship between visual lobe measurements and search performance has led to the suggestion that it may be possible to improve search efficiency through training to expand the visual lobe. Gramopadhye et al. (2002) found that a training protocol designed to increase visual lobe size produced positive transfer on a mock industrial inspection task, allowing trained subjects to detect faults on a simulated aircraft fuselage faster than untrained control subjects. Other researchers have suggested that training to expand the visual lobe may improve driving performance in older adults (Roenker et al. 2003).

Search Accuracy

Much of the previous discussion has focused on mechanisms that control search time. Of equal importance are processes that determine search accuracy and, in particular, target detection rates. Clearly the goal of visual scanning is to bring the target object within the visual lobe, near enough to the point of fixation to be recognized. By itself, however, a fixation on or near the target does not guarantee successful detection. In many tasks, a searcher may seem to gaze right at the target without detecting it. Examining the eye movement patterns of radiologists searching for pulmonary nodules in chest radiographs (e.g., Kundel, Nodine, and Carmody 1978) found that approximately 70 percent of false negative responses (i.e., misses) were the result of failure to recognize a target even after fixating directly on or very near it. A study of simulated baggage x-ray screening (McCarley, Kramer et al. 2004) found similarly that roughly 35 percent of targets went undetected even after they were fixated. Such lapses of recognition are most likely, of course, in tasks where the target is faint or camouflaged. Data also suggest that the likelihood of a failed recognition increases when the target stimuli are less familiar. Thus, even operators who are highly trained at searching for knives in a baggage screening task may have great difficulty recognizing a new target knife the first time it is encountered in a cluttered bag (McCarley and Carruth 2004; McCarley et al. 2004). On the basis of such results, Smith et al. (2005, p. 457) concluded that the processes responsible for recognizing and categorizing novel objects "suffer nearly a complete collapse when facing visual complexity and multi-item displays."

On trials where no target is present, the observer must continue scanning until reaching a decision that a no response is appropriate. Drury and Chi (1995) use the term *stopping policy* to denote the searcher's rule for ending a target-absent search. The standard serial and parallel models assume that a target-absent search will be exhaustive, with the operator executing a target-absent response after all items within the search field have been inspected once and classified as nontargets. In fact, human searchers often adopt more complex stopping policies. For example, searchers may limit their attentional scanning to items that are a relatively close match to the target object (Chun and Wolfe 1996) or, as the findings of Theeuwes (1996), discussed in chapter 4, indicated, may terminate search without inspecting locations that seem unlikely to contain a target. In either instance, a false negative response (i.e., a miss) can result because search terminates before reaching the target. In other cases, the operator may adopt a stopping policy more conservative than a conventional exhaustive search, taking time to double-check some items or locations before emitting a target-absent response (Cousineau and Shiffrin 2004). The importance of an appropriate stopping policy is illustrated by data from colonoscopy screening indicating a near-perfect correlation ($r = .90$) across doctors between mean polyp detection rates and mean search times for trials on which no lesion was detected; doctors who employed a more conservative stopping policy, taking longer on average to reach a no-polyp judgment, showed higher rates of successful polyp detection than those who tended to terminate search more quickly (Barclay et al. 2006). These data also indicate that a stopping policy can vary across different workers in a common task, showing consistent individual differences.

The stopping policy is also determined in part by the expected ease of target detection, by the payoffs associated with various types of correct or incorrect responses, and by the costs of errors relative to search time. Thus, for example, target-absent RTs increase when the target being sought is expected to be difficult to find (Drury and Chi 1995) and when the cost of a missed target grows larger (Chun and Wolfe 1996; Drury and Chi 1995). An especially powerful determinant of stopping policy, finally, is target frequency, which of course affects target expectancy. Wolfe, Horowitz, and Kenner (2005), for example, found that miss rates increased from 7 to 30 percent when the target frequency decreased from 50 to 1 percent. Such results suggest that one valuable way to improve target detection rates in tasks such as airport baggage screening, where true threats are encountered infrequently, is to introduce occasional mock targets (Wilkinson 1964), an approach that aviation security agencies have in fact begun to take.

Conclusion

Visual search is of great importance both inside the psychologist's laboratory as a way of testing theories of attention, and outside as an attentional component of many real-world user tasks. Importantly for application, search can be modeled as a mixture of parallel and serial processes. Designers strive to render targets amenable to parallel processing; to construct displays that guide search toward likely targets, thereby limiting the effective search field; and to instruct users on whatever spatial or contextual constraints typically exist.

Our understanding of search owes much to material covered in the previous chapters, but we also clearly see the relevance of much of this material to chapter 10 (i.e., training and individual differences) as well as to chapter 7 (i.e., effort), which revisits the issue of effort-related search termination. The next chapter turns from the largely serial aspects of visual search to the parallel-processing aspects of interpreting visual information once it is found: the role of attention in displays.

6

Spatial Attention and Displays

Introduction

Chapter 5 discussed the many characteristics of visual search, providing an important example of selective attention. That is, we could adopt the rough metaphor of spotlight, moving through space until the target object was selected or found. In this sense, much of the theorizing on visual search was consistent with the notion of space-based attention.

The current chapter does three things that build on this fundamental spatial metaphor. First, consideration is given to the implications of a space-based concept of attention to focused and divided attention as well as to selective attention. Second, space is contrasted with a second entity over which attention can be defined—the object, that is, object-based attention—by which is shown how objectness encourages parallel processing. Third, a description is given of applications of both space- and object-based attention theory to the important issue of designing displays, using the proximity compatibility principle as the underlying design framework for building displays that harmonize with the capacities of the human's limited attentional resources.

Space-Based Attention Theory

The metaphor of the spotlight, or flashlight, can be adopted to characterized space-based theory (Brown and LaBerge 1989; Wachtel 1967). Accordingly, the spotlight moves across the environment to attend to things in different locations. However, as the spotlight illuminates a target or other object of interest, its beam can only narrow so much as it zooms in to inspect. Hence, all visual information remaining within the beam, even if it is unwanted, will get processed, and that unwanted information will divert resources from processing the wanted information. That is, such information causes a disruption of focused attention on the wanted or relevant information. These characterize the problems of trying to drive while looking through a scratched or dirty windshield or examining a cluttered map to try to read detailed information.

The findings of focused attention disruption are well captured by the so-called flanker paradigm, developed by Eriksen and Hoffman (1973) as discussed in chapter 3, in which response time to classify a letter was disrupted by response-incompatible letters that closely flanked the target letter. But this disruption diminished as the flankers were moved outward, defining an area of the spotlight for mandatory processing that Broadbent (1982) estimated to be approximately one degree of visual angle. However, disruption of focused attention can occur across a wider range of space. For example, Mori and Hayashi (1995) observed disruption across somewhat larger visual angles from nearby but irrelevant computer windows as people were trying to process information from the relevant windows.

The previous example stressed the unwanted processing of response-incompatible stimulus information. This was, in part, because the degrading effects of these are starkly revealed in performance (i.e., response-time [RT] delays). However, it should be noted that nearly any stimulus items within this small region of visual angle around the wanted element can disrupt processing of the latter: the real-world effect of visual clutter.

Thus, although close spatial proximity can inhibit the focus of attention, so, intuitively, can more distant spatial separation inhibit divided attention between two visual sources (Wickens 1993; Wickens, Dixon, and Seppelt 2002). This divided attention cost to spatial separation does not appear to be linear but instead is represented by different components, as shown in Figure 4.5. Several references to that figure are made in this chapter, so it might be wise to bookmark it. Thus, when two sources are close together, there may be no cost to divided attention, but once they are separated by a few degrees of visual angle (both are not simultaneously within the useful field of view defined in chapter 5), then there is a cost associated with eye movements. Then, as the separation grows further and enters what is called the head field, requiring neck rotation to fixate on each of the two separated elements, an abrupt increase in cost is encountered. Divided attention, or dual-task, performance costs based on spatially separated sources follow a function corresponding to that shown in Figure 4.5 (Wickens, Dixon, and Seppelt 2002).

Object-Based Attention Theory

Compelling evidence now suggests that another dimension besides space can affect both focused and divided attention: whether or not an element B belongs to the same object as an attended element A. If B is to be ignored (i.e., is irrelevant) and belongs to the same object, processing of A will be hurt compared to the case when B belongs to a separate object. If, in contrast, B is also supposed to be processed (i.e., divided attention between A and B), its belongingness to the same object will help. Two examples of belonging to the same object are (1) the expression of gender and emotion of a perceived face; and (2) the location and size (representing population) of a circle representing a town on a map. In fact, there are actually three dimensions represented in the map example, since the location will represent both the X and Y coordinates.

An experiment by Duncan (1984) illustrates the benefits of belonging to the same object for divided attention. Duncan used stimuli, two examples of which are shown in Figure 6.1 with a larger or smaller box, having a gap on

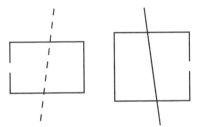

Figure 6.1 Two examples, each illustrative of the two objects, a rectangle and a slanted line, in Duncan's (1984) task.

one side or the other (one object). Each box contained a line (a second object) that was either dashed or solid and slanted left or right. Duncan found that judgments of two attributes (i.e., divided attention) were made better when both attributes belonged to the same object (e.g., box size and gap side) than when one belonged to each object (e.g., box size and line orientation). Importantly, the amount of visual scanning—separation by space—was equivalent in both conditions. In another experiment, Lappin (1967) found that subjects could report the size, shape, and color of one object just as well as they could report a single attribute of the single object—demonstrating perfect parallel processing between the three attributes of a single object—and far better than they could report either a single attribute of three objects (i.e., the shape of the three) or a different attribute on each of the three objects.

   In contrast to how a single object helps divided attention, some of the most compelling evidence for the cost of focusing attention on one feature of an object comes from the Stroop task (MacLeod 1992; Stroop 1992). Look at the set of words in Figure 6.2, and try to read, as fast as possible, the color of ink of the word (e.g., "white, gray, white, gray ...") and not the name of the word. Do this with the list on the left. Then do the same task (i.e., report ink color) with the symbols on the right, and notice how much more fluent is your reporting on the right (e.g., "gray, black, gray, white ..."). The difference is simply that each item in the list on the left is an object with two attributes: the word name and the color ink. When the name offers a response that is incompatible with the relevant attribute on which you should focus your attention (i.e., ink color), your response is slowed by the irrelevant element in exactly the same fashion that resulted with the Eriksen and Hoffman (1973) flanker task. Kahneman and Treisman (1984) integrated such evidence as we discussed already to propose their object file theory of attention, which postulates that perceptual processing is parallel within the features of a single object but is serial across different objects. Thus, when the ink color and semantic properties are processed in parallel, and they offer incompatible color responses, there is interference. Kahneman and Treisman's research showed that Stroop interference is greatly diluted or is eliminated altogether when the semantic properties are separated (i.e., belong to a different object) from the ink color.

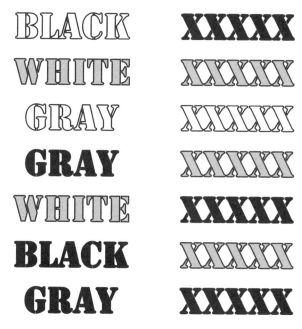

Figure 6.2 The Stroop task, usually performed with different colors, or hues, but here rendered in black, gray, and white. Going down the columns, report the color ink of each string, first on the left, then on the right.

The Proximity Compatibility Principle

At this point we can summarize the collective results of what we have discussed as shown in Figure 6.3. On the left, two or more elements on a display or in the natural environment can be either close to each other or distant, in which closeness can be defined by either spatial proximity or belonging to the same object. On the right, we can speak of tasks requiring either divided attention between those elements or focused attention on one while ignoring the other. We speak of the first, divided attention, as close task proximity since both display elements are needed and the second as distant or low-task proximity since only one is required and the other must be kept separate. There is then a compatible mapping between the display format on the left and the task requirement on the right as shown by the heavy arrows: close (high) proximity of the display supports close task proximity (divided attention), whereas low proximity (great separation) of the display supports low task proximity (focused attention). The dotted arrows show low-compatibility mappings. This relationship is rendered graphically at the bottom of the figure. Importantly, however, this relatively simple mapping, defining the proximity compatibility principle (Wickens and Carswell 1995) can be elaborated considerably on both the display proximity and the task proximity end, as shown in Figure 6.4 and as described in the following sections of this chapter in which this elaboration is integrated with real-world examples.

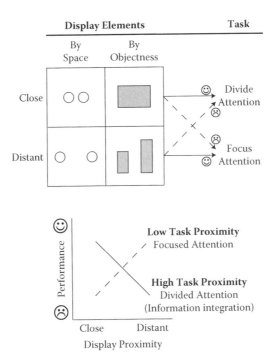

Figure 6.3 Conceptual representation of the proximity compatibility principle. At the top, different display proximities (left side) created by space or objectness are either compatibly (solid arrows) or incompatibly (dashed arrows) mapped to different attention tasks (right side). Below is shown a graphical depiction of the interaction expressed by the proximity compatibility principle, with good performance represented by higher values on the Y axis. Note that sometimes the term *task proximity* can be interchanged with *mental proximity*.

Task Proximity

In elaborating the attention task on the right side of Figure 6.3, it is important to distinguish focused attention from two different types of divided attention tasks. The first is dual-task processing, in which each display element is associated with a separate response and goal (e.g., looking at a map and looking at the car's heading on the roadway). The second is an *integration* task, in which attention must be divided between the two elements but in which both are mapped onto a single task (i.e., cognitive or motor response) and so their combined implications must be mentally integrated. Such integration tasks are common in the real world—for example, comparing one's generated answer with the correct answer on a key on a test question or comparing the geometric coding of a data point on a graph with the legend that says what that data point stands for. In the following discussion, the proximity compatibility principle is designed to most directly compare focused attention with divided attention integration tasks, since dual-task situations, though having

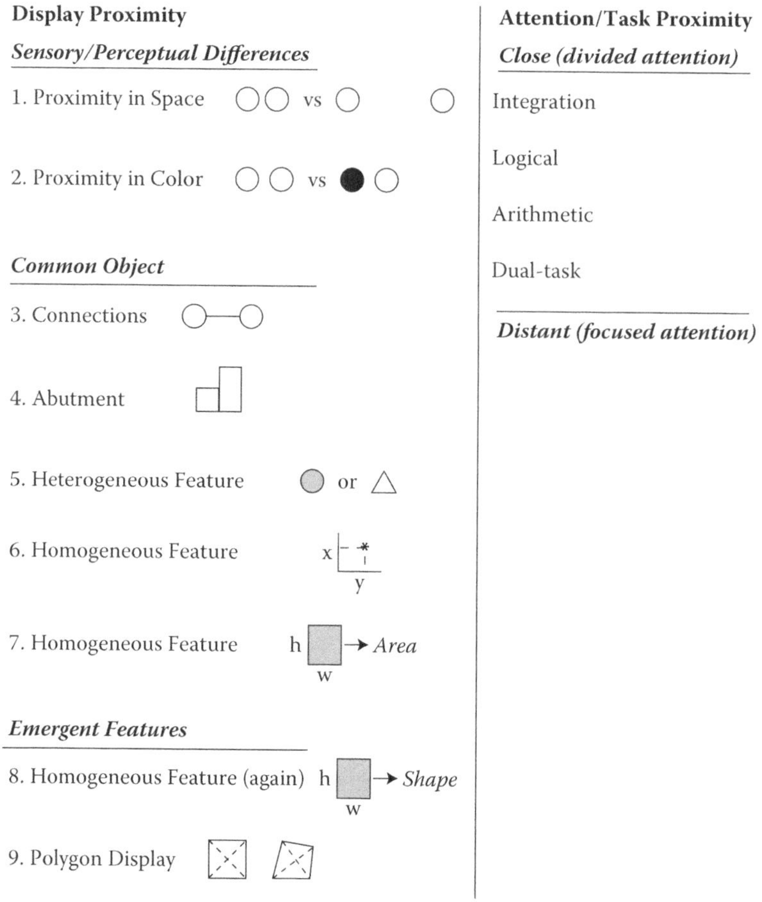

Display Proximity

Sensory/Perceptual Differences

1. Proximity in Space   ○○ vs ○   ○

2. Proximity in Color   ○ ○ vs ● ○

Common Object

3. Connections   ○—○

4. Abutment

5. Heterogeneous Feature   ● or △

6. Homogeneous Feature   x ⊢ *

                         y

7. Homogeneous Feature   h ▢ → *Area*
                            w

Emergent Features

8. Homogeneous Feature (again) h ▢ → *Shape*
                                  w

9. Polygon Display

Attention/Task Proximity

Close (divided attention)

Integration

Logical

Arithmetic

Dual-task

Distant (focused attention)

Figure 6.4 The dimensions of display proximity (left side) and task proximity (right side) as defined in the text. Most examples on the left side are of high display proximity, so in the framework of Figure 6.3, they would be compatibly mapped to the integration tasks on the right. The different rows are described in the text.

commonality with integration tasks, also share some features with focused attention (Carswell and Wickens 1996). Divided attention in dual tasking is discussed extensively in the next two chapters.

Display Proximity

Now, turning to the left side of Figure 6.3 and particularly Figure 6.4—the display or perceptual proximity side of the figures—this concept can be defined in several ways that elaborate and extend on the two primary categories of space- and object-based attention. Each of these ways are represented by different rows in Figure 6.4.

Sensory and Perceptual Similarities: Proximity in Space and Color

As discussed already, space-based proximity is related strongly to the effort required to move attention from one location to another (Figure 4.5, Figure 6.4, row 1). A classic example is the book designer's goal to keep the figure on the same page as the text that discusses that figure rather than to require a page turn to go from the figure to the text. Such an information-access task—page turning and text search—competes for cognitive resources with the retention of information relevant to one source (e.g., figure or text) while accessing the second as required in cognitive integration (Liu and Wickens 1992a). Dupont and Bestgen (2006) applied this design guideline of proximity to embedding icons within the text right next to the text description. Such embedding yields proximity between icons and text and was found to help users more effectively program a display that also contains the icons.

It is important to realize that spatial distance, or difference, is only one way to create a sensory and perceptual experience of closeness between two elements (Garner 1974). These elements can also be the same color (Yeh and Wickens 2001b), including both hue and intensity (Wickens, Alexander, et al. 2004) as shown in row 2 of Figure 6.4. These commonalities, like spatial proximity, will also make it easier for people to divide attention between them or to integrate them in an otherwise cluttered visual field. Such a finding relates back to material on visual search discussed in chapter 5: When two target elements share a common preattentively processed feature (e.g., color), both will pop out from the background and therefore will be more easily related or compared, or integrated.

As a concrete example, this use of color has been suggested as an aid to air-traffic controllers in examining a display of several aircraft at many different altitudes while the controllers are trying to integrate or to understand the joint trajectory of two particular aircraft that may be on a collision course and therefore at the same altitude. Two same-color techniques can be employed to help such mental integration. First, all aircraft flying at a given altitude can appear in the same color (Remington et al. 2001), making it easier to direct attention to and divide attention between, or to integrate, those that may represent potential collision threats. Second, a particular aircraft pair on a conflict trajectory could be colored red, thus making it easier for the controller to both notice these and to understand their joint trajectory, the latter being an integration task. Such a concept has been employed in an air-traffic control system called the user request evaluation tool (URET) (Wickens et al. 1998; see also Figure 6.5). Note that from a designer's standpoint, close display proximity can be created by using the common color when it is otherwise impossible to relate the two aircraft by moving them closer together in space (row 1) since their position in space is determined by what the aircraft are doing and not by what the designer wants. In fact, this constraint is common to all maps; the designer simply cannot move things around on a map to make it easier for the user to relate them.

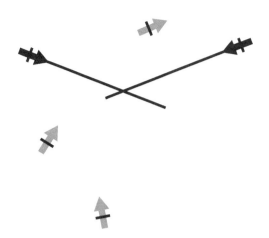

Figure 6.5 A schematic example of the URET designed to help the air-traffic controller understand the joint-conflict trajectories (an integration task) of two aircraft that may be on a collision course. Note that the two conflicting aircraft are both joined by lines (linking) and are represented in a common color (highlighting).

Object Integration

As we move down to rows 3–7 on the left side of Figure 6.4, we see that objectness can be defined in several ways. First, larger objects can be created from smaller ones by connecting the contours of two smaller objects (row 3, the two circles on the barbell) or by abutting them in space (row 4). Second, objects can be defined by the separate attributes of a single blob such as the color, brightness, and size of a geometric shape (row 5), the X–Y position of a point in a graph (row 6), or the height and width of a rectangle (row 7). We now elaborate on each of these aspects of object-based proximity in turn.

Connections and Abutment

Just as common color can relate two otherwise spatially separated objects, so it is possible to do so with a line or link that joins them (row 3), as attention appears to be relatively automatically drawn along connecting line features (Jolicoeur and Ingleton 1991). Thus, note in Figure 6.5 that in the URET display the two conflict airplanes are not only common colored but are also joined by a line. As another example, in presenting complex device instructions that combine pictures of the device components with sentences describing what to do (e.g., "turn the knob to the left") large gains in usability are achieved when the printed instruction sentence is linked to the pictorial rendering of the actual knob by a dashed line (Tindall-Ford, Chandler, and Sweller 1997). Although such links can create a certain amount of clutter on a display—as we shall see, possibly disrupting focused attention—this clutter can be minimized by keeping the intensity contrast of the linking lines with the background relatively low. It is possible to select contrasts high enough so the links can be perceived but not so high as to disrupt focused attention on the connected elements (Wickens, Alexander, et al. 2004).

Heterogeneous- versus Homogeneous-Featured Objects
Row 5 of Figure 6.4 shows two objects created by three heterogeneous features: size, brightness, and shape. Such features are described as heterogeneous because they are found to be processed by different perceptual analyzers (Treisman 1986). This is typical of the data point in many graphs (e.g., black square, white square, black circle, white circle) or of the representation of a town on a demographic map (e.g., population, political leaning, mean income represented by size, color and shape). The reader will recognize that the stimuli used in the Stroop task shown in Figure 6.2 are heterogeneous objects, with a semantic and a color dimension. Rows 6 and 7 show two homogeneous-featured objects, each object defined by its height and width. These are said to be homogeneous because a single perceptual analyzer—of spatial distance—defines both. A display designer confronted with the question of how to represent two aspects of a single entity may be challenged as to whether to use heterogeneous or homogeneous features. The answer appears to lie in both the kind and degree of integration or task proximity that is required (Carswell and Wickens 1996; Wickens and Carswell 1995):

- If the integration goal is simply to have the user consider both aspects of the entity at once in a Boolean logical operation (e.g., Is a given city both large and conservative in population?), then heterogeneous-featured objects are the ideal choice because they best allow parallel processing of the two dimensions (Lappin 1967)—some would say this supports divided attention between the dimensions. In particular, heterogeneous features of objects are a very economical way of presenting lots of information on a space containing several objects (e.g., map with several cities) because all attributes of the single object can be processed in parallel, the processing being divided between the different analyzers. Thus, they are good clutter reducers. Heterogeneous-featured objects also effectively support redundancy gain, in which all features lead to the common response: Consider the stop sign with color (e.g., red), shape (e.g., octagon), and semantic meaning (e.g., stop) being the prototype of a redundant heterogeneous-featured object display.
- If the integration goal is an arithmetic or comparative one, however, heterogeneous features no longer provide the same benefit, since each feature is expressed in its own different perceptual currency, which cannot easily be compared. For example, an aircraft pilot who wants to compare actual speed with target speed does not want one to be expressed on a spatial dimension and the other in a color-coded dimension. Instead, it is better for both to be spatial, perhaps as the height of two connected bar graphs (row 4). Many integration tasks

involve mental multiplication where the rate of some operation (e.g., rate of travel, rate of spending) is to be multiplied by the time of operation to produce a total quantity measure (e.g., distance traveled, total amount spent). Here heterogeneous features poorly serve multiplication, but homogeneous features do so better; this is particularly true for the height and width of a rectangle (row 7), in which the area of the rectangle display is directly equal to the product of the two variables (Barnett and Wickens 1988). Thus, a cognitively computed quantity, or product, becomes directly replaced by a perceptual one. Since perception often proceeds more automatically than cognition, this replacement is a desirable human factors goal and is one basic underpinning to the design of ecological displays that is discussed later in this chapter (Vicente 2002; Vicente and Rasmussen 1992). The user does not have to multiply numbers in his or her head but simply and directly can perceive the size of the rectangle.

Proximity by Emergent Features

An emergent feature of a display is a new perceptual property that emerges from the combination of its single-dimension elements if these elements are combined or configured in a particular way (e.g., into a single object) that would not exist if they are presented separately (Pomerantz and Pristach 1989). We have just seen an example of such an emergent feature in terms of the product of height and width of a rectangle emerging to yield a directly perceived feature: the area (row 7). Correspondingly, the difference between height and width in this display produces a second emergent feature of shape, as also shown in row 7: A tall, skinny rectangle looks very different than a short fat one, even as the two may have the same area.

To be useful in display design, emergent features must be mapped to the mental quantity that is to be integrated (Bennett and Flach 1992). We can illustrate this by describing a medical display for monitoring patient respiration developed by Cole (1986). Two rectangles displays are presented side by side as shown back in row 4 of Figure 6.4, one showing the natural breathing of the patient and the other showing the artificial breathing imposed by the respirator. Such displays are coded so that the height of the rectangle represents the depth of breathing (i.e., amount of oxygen supplied on each breath) and the width represents the rate of breathing (i.e., breaths per minute). Thus, the total amount of oxygen is represented by the height × width = area (emergent feature 1). A quick glance can identify the relative size of the two rectangles and therefore the amount of work the patient is doing compared with the respirator; these measures of respiratory work are more easily integrated because they are connected by abutment. Furthermore, the style of patient breathing (i.e., short panting versus slow, deep breaths) can be rapidly perceived by the shape of the rectangle (emergent feature 2). Both relative amount and style—each integration tasks dependent on dividing

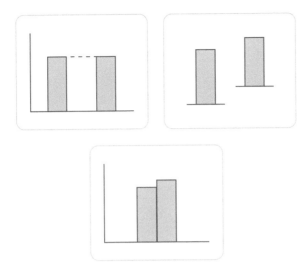

Figure 6.6   The role of common baseline alignment in creating an emergent feature (left side) to signal equal operating parameters. This is not available on the right side because the baselines are not aligned. The effectiveness of this emergent feature can be amplified when close spatial proximity or abutment is used (bottom row, illustrating nonequal parameters).

attention between breathing rate and depth—can be easily discerned by a quick glance to perceive these two emergent features: size and shape. Such relatively automatic perception could not be accomplished if either of the basic breathing variables were presented separately or were represented by heterogeneous display dimensions.

Importantly, research and intuition have suggested that emergent features need not be created by an object display (Sanderson et al. 1989), although they are usually encouraged by objectness. Figure 6.6 provides an example of two bar graphs (separate objects), each perhaps representing the desired and actual temperature of some operation and supporting the divided attention integration task of assessing that the two values (desired = actual) agree. To the left, it is easy to perceive that the system is operating normally: the height of the two bar graphs is identical. To the right, the same integration judgment—judging that the difference in height is zero—is considerably more difficult. The reason is that aligning the two bars to a common baseline produces an emergent feature on the left: the colinearity of the tops, which automatically signals equivalence. It is as if one could imagine a ruler laid flat across the top, shown by the dashed line. One also might notice in Figure 6.6 how different display features of similarity can work in conjunction: If the same-baseline bar graphs are also close together or, particularly if they are abutted, creating a single object as shown at the bottom this emergent feature of colinearity is extremely salient since its absence will be

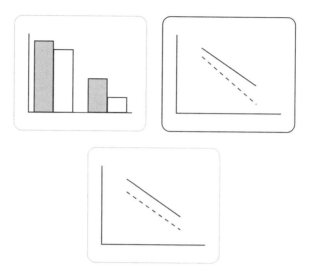

Figure 6.7 Contrasts the ease of understanding line graphs (right panel above) with bar graphs (left panel above), when a complex interaction relationship exists between the two variables. On the right, the diverging angles of lines (an emergent feature) signals the interaction. The value of this representation increases as the number of variables increase (Figure 6.9). In the line graph below, an additive relation between the two independent variables is signaled by the emergent feature of parallel lines.

signaled by the break in the line across the top. This is a sensory and perceptual feature (vernier acuity) to which humans are extremely sensitive.

Another important emergent feature is the slope of a line that may connect two objects in a graph. For example, consider the graphs in Figure 6.7. On the left, it is possible to note that bar graphs are of different height or that four bar graphs create an interaction between two variables in a 2 × 2 experimental design. However, when these Y-axis data points are connected by lines to create a line graph as shown on the right, then the height differences become very salient as represented by the slope of the lines connecting the two points. This slope is an emergent feature. Furthermore, the different magnitude of the effect, or interaction, is expressed visually by the emergent feature of the angle between the two lines. Indeed, when the two variables are additive and the effect of one is the same size at both levels of the other, the parallel aspect of the two lines also becomes an important emergent feature. This is shown at the bottom.

Before we leave the discussion of emergent features, we note one additional emergent feature that can often be created and exploited in display design: the symmetry of an object or configuration. The property of symmetry is one to which human perception is naturally tuned (Garner 1974; Pomerantz and Pristach 1989), so if a symmetrical configuration can be directly mapped to a critically important display state, then a well-conceived emergent-features display will be achieved. An example often cited is the polygon display (Beringer and Chrisman 1991; Woods, Wise, and Hanes 1981) such as that shown in row 9 of Figure 6.4. Here the normal operating levels of four parameters of a system are represented by a fixed, and constant, length of the sides of the quadrahedron—or the length of four radii from the center. When all four are at this normal level, a perfect square is shown, as on the left. The square is both vertically and horizontally symmetric and easily perceived as a square. When any variable departs from normality, symmetry is broken, and the distortion is rapidly noted, as shown to the right.

In closing our discussion of information integration, we note two important aspects of emergent-, or homogeneous-, featured object displays. First, the creation of such displays involves considerable creativity on the part of the designer—some would say as much art as science—not only to identify the critical system states to be perceived by the display user (i.e., the integration task mapping through cognitive task analysis) but also to think of ways that this mental integration can be best supported by creative configuration of the display dimensions. The second aspect is an obvious point that may have already been noted by the reader: Suppose the display user does not care about the emergent integration quantity but instead needs to know the precise value of a particular underlying dimension (e.g., What is the patient's rate of breathing?). Would this focused-attention task be hurt by an object rendering? This question of course brings about the issue of focused attention and the extent to which design features that support successful divided attention for integration may actually harm focused attention. If so, then this represents the so-called no-free-lunch effect to which we now turn.

Costs of Focused Attention: Is There a Free Lunch?

The proximity compatibility principle analytically proposes that there will be an interaction between display and task proximity, graphically depicted at the bottom of Figure 6.3. In its purest form, it predicts that closer display proximity—however achieved (Figure 6.4)—will improve performance on integration tasks and will disrupt performance on focused-attention tasks. However, more modest forms of the interaction are also consistent with the principle, namely, that increasing display proximity, though helping integration tasks, may not help or will help less the performance of a task when focused attention is required on one of its displayed elements even if such close display proximity will not actually hurt focused attention. By now, the negative effects or diminished benefits of high-task proximity on focused attention are well documented (Wickens and Carswell 1995), even if this effect is typically smaller in its magnitude compared to the benefits of close display proximity for integration (Bennett and Flach 1992). The following are two examples of negative effects of close display proximity.

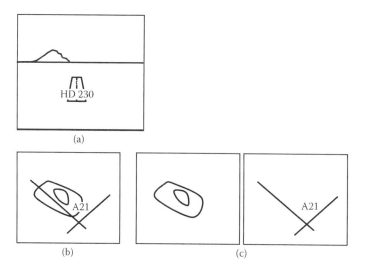

Figure 6.8 (a) Schematic example of a HUD, where the pilot can see the runway beyond and the horizon and mountains while looking through a windshield display of superimposed digital information. Close display proximity of the near and far domain is thus created. (b) Two overlaid databases, one of linear air routes and the other of a severe weather contour. (c) The same two databases of (b), shown in two different displays.

Example 1

Although reducing the separation between two elements improves the ability to compare them (integrate) particularly in a cluttered display, bringing them so close together that they overlap will certainly disrupt the ability to read one, if the other is to be ignored, imposing a major source of clutter. This can best be seen with the head-up display (HUD), which superimposes a display over the outside field of view for the aircraft pilot or vehicle driver (Figure 6.8a) (Fadden, Ververs, and Wickens 1998, 2001; Wickens, Ververs, and Fadden 2004). On the one hand, such superimposition can help tasks in which the HUD information needs to be compared, or integrated, with the outside world (e.g., Does my HUD guidance line up with the true runway on which I am about to land?). On the other hand, the overlapping clutter can sometimes hinder the ability to read particular pieces of information on the HUD or, in particular, to see unexpected elements only visible in the outside world through the HUD, such as a vehicle lurking on the runway (Fadden, Ververs, and Wickens 1998, 2001). Both of these are focused-attention tasks. A corresponding set of effects to those of the HUD are produced by database overlay (Kroft and Wickens 2003; Wickens 2000), as depicted in Figure 6.8b. When a weather map is overlaid on a flight-plan map (i.e., close display proximity), it certainly helps the viewer understand if the bad weather is on the flight path, which is a divided attention–integration task, but the cluttered weather overlay will hinder the ability to read air-route labels (i.e., focused attention) compared with a layout with separated side-by-side displays of the two databases (Figure 6.8c).

Example 2
Configuring two quantitative variables as points in an X–Y space (Figure 6.4, row 6) (Goettl, Wickens, and Kramer 1991) may help the viewer to understand the joint implications of, or corelation between, X and Y (integration) but will hurt the ability to read (i.e., focus attention on) the exact value of each, as now this value needs to be projected or visually extrapolated to an adjacent X or Y axis. An alternative here is to provide an added (e.g., digital) display of exact value.

In considering these and other examples, it is important to understand that there are several circumstances in which closer display proximity to aid integration, though not helping focused attention, does not hurt it either. For example, as noted before, with regard to spatial separation such costs to focused attention only emerge when spatial separation is decreased below about one degree of visual angle and are then amplified when true overlap occurs. But as we saw earlier in this chapter—and in chapter 4 (Figure 4.5)—the costs of increased separation to divided attention are relatively continuous across a wide range of angles above one degree. Thus, decreasing separation, say, from twenty to two degrees will help integration and will not hurt focused attention. As another example, rendering two items in a cluttered display the same color (or intensity) will not hurt the focus of attention on either—and indeed may even improve focused attention performance if this unique color coding of the pair allows a visual searcher to more rapidly find a single one against a background of many other items (Wickens, Alexander, et al. 2004). In the same vein, using a line to connect two elements (e.g., a line connecting two dots on a line graph will little hinder the ability to extrapolate the line's position to the axis, a focused-attention task, compared with a bar graph, even as that line, producing an emergent feature (line slope) will improve the ability to note the difference in the two points' values (the integration task). In short, sometimes there is a free lunch—or at least a cheap one—if proximity is used with care. Thus, a designer who must configure a display to support an array of focused and integration tasks may, by careful selection of different proximity metrics, be able to support the best of both worlds. An example is the application of the proximity compatibility principle to graph design, which we discuss next.

Applications to Graph Design

The proximity compatibility principle is directly applicable to designing effective graphs that the user can process without investing unnecessary cognitive effort, although many principles of good graphics go well beyond the attention principles embodied by this principle (Gillan et al. 1998; Kosslyn 1994; Wickens and Hollands 2000). Consider the graph in Figure 6.9, which

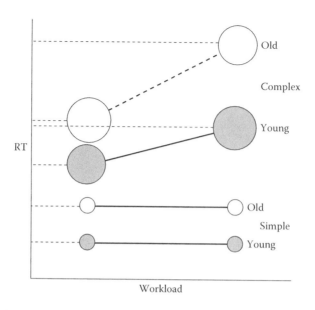

Figure 6.9 Application of several aspects of the proximity compatibility principle to the design and layout of graphs, in this case the graph of the results of a 2 × 2 × 2 experimental design.

depicts the hypothetical results of an experiment in which the combined effects of workload, task complexity, and age on response time are portrayed. Five features in the design of this graph, relating to the features of proximity in Figure 6.4, adhere to the proximity compatibility principle:

(1) Object integration: Note the heterogeneous-featured object that makes up each data point, representing age (intensity) and task complexity (symbol size). There is also a redundant object integration since the link between the two data points within each age group is coded by line type (dashed, solid).

(2) Connections: The links, or connections, between data points reveal the workload effect—and what moderates it—in terms of the emergent feature, line slope, as Figure 6.7 shows. For example, this feature helps integration so that attention can be focused on the large dots to see how workload affects performance with complex tasks or on the white dots to see how workload affects older participants. Or to obtain full integration, one can easily see that workload had no effect with the simple task (two parallel lines), but with the complex task workload increases RT and does so disproportionately more for older people (the diverging angle between the two complex lines).

(3) Spatial proximity of legends: The legends are placed directly next to the lines they label. These two elements—legends and lines—need to be related or integrated in the mind as the reader considers what each

line represents. There is much less cognitive effort imposed by this integration task than would be the case if the legends were placed in a separate legend box.

(4) Spatial proximity of lines: All four lines have been given high proximity by plotting them on a single panel. This display integration allows the three-way integration described in (2) to be easily visualized. This would not be easy if the simple task data—or the data of old people—were plotted in one graph and the complex task data—or the data of young people—were plotted in another graph (i.e., display separation).

(5) Focused attention: Although the focus of attention on particular values for a single data point might be more inhibited here than if a table been used, note that the light horizontal lines connecting data points to the X axis support this task, yet their low intensity does not hamper the attention to the overall integration, yielded by the line slopes.

The graph in Figure 6.9 was meant to emphasize how the workload effect was moderated by age and complexity; hence, the most important mental or task integration—workload effect—was represented by the line and thus was supported by the emergent feature of slope. If instead, the most important point of emphasis is the age effect, then age rather than workload would be represented as the single object line.

Our example for applications of the proximity compatibility principle to graph design is actually a fairly simple one: eight data points, three independent variables, and one dependent variable. However, the principle and its guidance for the control, direction, and support for attention is very applicable for much more complex data visualizations (e.g., Wickens and Hollands 2000; Wickens, Merwin, and Lin 1994), here there may be massive amounts of quantitative data, with hundreds of data points (objects) defined on scores of different dimensions (North 2006; Tufte 2001). The ultimate challenge in designing such visualizations is to help support the user to integrate information across objects or dimensions to gain insight and understanding into relations that are not yet known. The previous examples have shown how display similarity and proximity can support such integration in simple graphs; such techniques can be extended in many creative ways to the more complex graphs supporting visualization (Liu and Wickens 1992b; North 2006; Wainer and Thissen 1981).

Ecological Interfaces, Navigation, and Supervisory Displays

Our discussion of the proximity compatibility principle and attention to multielement displays leads us to the concept of ecological interface design (EID), a design process that supports the operator in maintaining awareness of complex dynamic systems (Burns and Hajdukiewicz 2004; Smith, Bennett, and Stone 2007; Vicente 2002; Vicente and Rasmussen 1992). These might describe the needs of supervisors in the process control industry, the

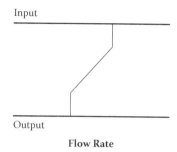

Flow Rate

Figure 6.10   An example of an ecological display, showing the imbalance between two quantities (e.g., flow in versus flow out of a reservoir). This rightward slope would indicate that the reservoir is filling: Output is input is greater than output. Desired steady-state behavior would be indicated by a vertical line connecting the two scales (adapted from Smith and Bennett 2006).

air-traffic controller, or the anesthesiologist. EID has many facets, not all of which are closely related to attention, but the following section highlights those that do.

EID and Emergent Features

The concept of *ecology* in EID describes the process of representing elements perceptually in a way that they are expressed in the natural world—and, it is assumed, within the mental model of the well-trained expert (Smith, Bennett, and Stone 2007; Vicente and Rasmussen 1992). To do so, designers focus on analog displays because most of the natural world behaves in an analog fashion, reflecting the laws and constraints of physics. Furthermore, displays are configurable in a way that highlights key relationships in the real world and represents these by emergent features within the display (Bennett, Toms, and Woods 1993). Clearly, such a philosophy discourages digital and symbolic displays, but it also discourages certain types of analog displays like round dials—unless these are explicitly representing a circular variable like a compass heading.

   As an example, in many systems there is a need for the system to preserve a balance to be operating normally. For example, the flow into a chemical or heating tank should equal the flow out of a tank, or in nuclear power facilities it is critical that the balance between mass and energy be preserved. Such a balance can be directly and intuitively perceived by the display form in Figure 6.10. This representation relates directly to our discussions of attention because of the degree to which the single object supports the parallel processing of its attributes. Other examples from EID relate to object-based safety parameter displays such as those described previously in Figure 6.4, row 9 (Woods, Wise, and Hanes 1981).

Integration over Space and Time

Systems in the process control industry may contain hundreds of different dynamic parameters to be displayed, clearly presenting a visual overload for a single workstation or computer display (Moray 1997). Such a situation will force selective attention to operate somewhat sequentially—even as emergent features previously described may support some parallel processing. However, good design can work to minimize the effort of information access across the various display panels. One way to achieve this is to minimize the need to view screens separated by time (i.e., sequentially) so that working memory load is reduced as a source of interference and so that these memory demands will compete less with cognitive resources required for navigation between screens. However, simultaneous views of multiple screens can so reduce the resolution of information on each screen that it becomes illegible (Kroft and Wickens 2003) even as information access costs are reduced. Effective creation of integrated ecological displays, however, can provide the benefits of simultaneous viewing without imposing major costs to display resolution (Smith, Bennett, and Stone 2007). In a process control simulation, Burns (2000) compared the three philosophies of sequential displays, simultaneous displays, and simultaneous-integrated displays as ways of representing the same process control information, with the simultaneous-integrated display incorporating the principles of EID. Burns compellingly showed the advantage of the latter when integration tasks (i.e., fault diagnosis) are required.

Visual Momentum and Navigation

When multiple views of a complex industrial system process, database structure, or a physical, or geographical, space are presented to the viewer who tries to integrate information between such views, the issue of how to support movement of attention while relating different views becomes critical. As a simple example, when you are in an unfamiliar territory, maintaining geographical awareness often involves relating—integrating—the layout presented on a map with the visual image looking forward (Wickens 1999b; Wickens, Vincow, and Yeh 2005). In this example as well as in many others, such an endeavor is cognitively described by relating a more global view of the entire space with a more local view of a particular region of the space.

In support of this endeavor to move attention between different but related or overlapping representations of a space, the technique of visual momentum has proven quite effective (Aretz 1991; Olmos, Wickens, and Chudy 2000; Wickens 1993; Woods 1984). This technique is one originally used by filmmakers to provide the viewing audience with a graceful cognitive transition between different cuts of the same scene, such as zooming and panning (Hochberg and Brooks 1978). In the case of industrial system monitoring, one application of visual momentum as shown in Figure 6.11a might be to pro-

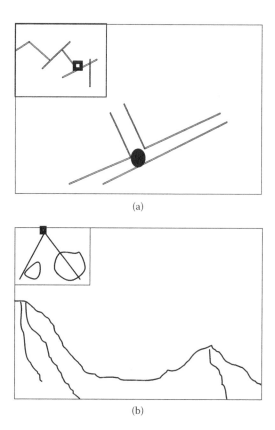

(a)

(b)

Figure 6.11 Two examples of visual momentum. In (a) a schematic representation of the piping of a full system is shown in the inset, highlighting the location of the particular valve rendered in the main part of the display. In (b) the pilot is flying southbound, toward the gap between two mountains, as shown in the global view inset. The field of view of the local view, shown in the main part of the display is rendered by the inverted wedge in the small global view. In both examples these features (i.e., highlight, wedge) help to show how the local view fits into the big picture of the global view and thereby helps the user to divide or switch attention between the two views as integration between the two views is required.

vide two views: a global view of the entire plant layout and a simultaneous zoom-in local view of a particular region of the plant where a problem has developed. To create visual momentum, the elements of the local view are highlighted on the global view, so it is easy for the viewer to see how one relates to the other and to move attention back and forth between them.

One visual momentum technique that has proven to be particularly valuable in navigating through real or virtual spaces is the wedge shown in Figure 6.11b, which shows how the information in the large forward view local display (shown below), is represented in the global map (shown as the inset above) by presenting the field of view of the former as a wedge overlaid on the latter (Aretz 1991; Olmos, Liang, and Wickens 1997; Wickens, Thomas, and Young, 2000). Such a technique of attentional support is particularly advantageous when one is looking or traveling in a direction that is opposite the normal orientation of the global view. This is often the situation with a north-up map when traveling southward. Levine (1982) (also see Wickens and Hollands 2000) showed how this can be applied to you-are-here maps, commonly found in malls and city areas.

Conclusions

In the interfaces with which we work, displays are usually visual and almost always contain multiple elements. Hence, it is not surprising that visual attention is challenged. Because of this challenge, it is the designer's goal to either foster as much parallel processing as possible or at least to reduce the effort of switching and selective attention, which is required to visually access related elements in sequence. The proximity compatibility principle, defining components related to both space- and object-based theories of attention, and related elements of proximity in EIDs represent ways that this can be accomplished. Such techniques then often help the sequential access of information. Chapter 4 described the important concept of information access effort, and the next chapter does so as well. Importantly, sometimes the display environment confronted by the human contains nonvisual elements as well: speech, tones, and even tactile stimulation. These multimodal aspects of display design are described in chapter 8.

7

Resources and Effort

The common exhortations *try harder* and *pay closer attention* are closely related. Both phrases appear to invoke the concept of effort: mental effort in the latter case and either mental or physical effort in the former. The deployment of mental effort can in turn be considered an exercise of attention. Kahneman (1973), in his classic book *Attention and Effort*, conceived of effort as a sort of mental energy or resources (Hockey, Gailliard, and Coles 1986). As a continuously allottable resource, it can be mobilized or demanded in continuously varying quantity. In this respect, the notion of attention as a resource contrasts markedly with the metaphor of a two-state (i.e., open or closed) bottleneck or the discrete attention switch (Tsang 2006). The concept of effort in psychology has also been tied closely to physiological characteristics of arousal through the autonomic nervous system (Gopher and Sanders 1985; Hockey 1997), although the two concepts are not synonymous.

As effort pertains to task performance, it can be invoked in two different ways: relating to the person and to the task (Kahneman 1973). First, the person can be said to invest varying levels of effort (i.e., the concept of try harder), reflecting a kind of strategy. Second, the task can be said to demand varying levels of effort to achieve a particular level of performance. These two contexts for effort as it relates to performance will both be seen to be reflected in both the concepts of mental workload and of automaticity, as discussed in this chapter.

Finally, we see that both person and task concepts of effort are equally relevant to both single- and dual-task performance, a distinction made in the following two sections. The chapter concludes with a section addressing the role of effort in learning.

Effort in Single-Task Choice

The general framework for considering effort in a single-task context is that the investment of excessive mental effort, like excessive physical effort, generally produces an unpleasant state that is to be avoided. Hence, people tend to be inherently effort conserving, particularly when placed in highly demanding environments, where attention theory is generally most applicable to real-world problems. Shugan (1980) wrote about the cost of thinking. With this effort conservation in mind, Figure 7.1 presents a very general model of effort in choice. Here, a person is confronted with a choice between two decision-making options. One requires little effort to carry out but may provide little value or an expected loss; it is risky. The second requires a

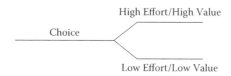

Figure 7.1 A generic model of effort and choice.

higher investment of effort but promises a greater value at the end—"no pain, no gain," as the saying goes. We can apply this simple representation to a number of different kinds of single-task choices.

Effort in Selective Attention: The Choices to Stop Search and Limit Surveillance

Chapter 5 discussed visual search tasks in which people were looking for something. Serial search can be considered effortful—assuming that more time and more eye movements translate to more effort. A critical issue in understanding search performance, as discussed at the end of that chapter, is stopping policy: How does an operator decide to stop searching and conclude that the target is absent? Very often, a target-absent response is executed at the time that the effort required to continue searching exceeds the expected value of finding the target—or, more properly, exceeds the expected loss of failing to find it. Thus, in many real-world environments, when the searcher reaches some total expended effort threshold search may be terminated early even if the target has not been found (e.g., "It's not worth searching anymore"). In Figure 7.1, this decision to stop searching corresponds to taking the bottom path. This threshold for the decision obviously varies for a number of reasons, just as the expected cost of not finding something varies with the value of the unfound object. Importantly, the effort concept here can be applied to a number of real-world searches, such as the decision to terminate a document search (MacGregor, Fischhoff, and Blackshaw 1987), to give up searching for a particular experimental phenomenon in the laboratory or, to place the issue in a recent context, to abandon the search for weapons of mass destruction in Iraq.

In most of the aforementioned contexts, effort can be expressed fairly directly in terms of time (Gray et al. 2006) and sometimes in financial resources expended; the two can be used interchangeably if time is money. However, as this chapter shows, time is not always equivalent to effort.

A different role of effort in visual search is applied to the surveillance task, in which one is no longer searching in a self-terminating manner for a specified target but rather is monitoring for a particular event—and monitoring will continue even after an event is detected. This may involve searching for enemy targets (Yeh, Wickens, and Brandenburg 2003; Yeh, Wickens, and Seagull 1999), for example, or monitoring an air-traffic display for changes in traffic heading or altitude that produce a conflict (Remington et al. 2001; Stelzer and Wickens 2006). Here effort may limit the spatial extent of scanning surveillance. If long glances (i.e., big saccades, head movements) are needed to check the periphery of a display, the effort-conserving searcher may focus search more heavily toward the center of the display and avoid the edges, a phenomenon known as the edge effect (Parasuraman 1986; Stelzer and Wickens 2006). Certain design features of the surveillance environment may inhibit peripheral searches more because they are effortful. For example, wearing heavy head-mounted displays, which would require effortful head movement, may keep surveillance more centralized than desirable (Seagull and Gopher 1997; Yeh, Wickens, and Seagull 1999). Providing panning features on a an immersive three-dimensional imaging device may lead users to focus more of their attention on the display shown by the default setting of the imager (i.e., straight ahead) and may fail to pan adequately to the sides (Wickens, Thomas, and Young 2000) even when targets of importance may appear at the sides. In this case the effort is what is associated with manual activity to control the panning device rather than head movement. Collectively, we refer to this as information-access effort—its role in visual scanning was discussed in chapter 4—in the context of the second letter of the SEEV model of selective attention.

Effort in the Choice to Behave Safely

In the previous section, the choice involved was really a choice between continuing or stopping a search. In many other situations, effort plays a role in the choice between engaging in safe or unsafe behavior, often at the workplace. Here we consider the two limbs of Figure 7.1 again. The lower limb is the choice to behave unsafely, which may have a negative expected value if an accident could result; consider, for example, failing to wear protective goggles or a safety helmet. The upper limb, which is safer (i.e., higher expected value), is also penalized by the effort imposed by a *cost of compliance* (Wickens, Lee, et al. 2004; Wogalter and Laughery 2006). This cost is often one that can be expressed in terms of the effort—here again, often time—required to comply (e.g., locating and putting on the safety equipment, taking the time to read and understand the safety instructions), although the cost also may incorporate a noneffort influence such as the discomfort of wearing the safety equipment. The cost may involve mental effort if, for example, the operator is required to read and understand poorly worded instructions for following safety procedures. Under such circumstances, this reading will often be bypassed.

Effort in the Choice between Decision Strategies

While the aforementioned examples described the influence of effort in dissuading an operator from behaving in a particular way (e.g., continuing searching, following safety procedures), a third important application of effort in choice is the role that it plays in guiding the choice between two different strategies of deciding between options. This application is based on a long line of research on decision making by Bettman, Payne, and Johnson and their colleagues, with scores of papers by the three authors in various order (see Bettman, Johnson, and Payne 1990; Johnson and Payne 1985; Payne, Bettman, and Johnson 1993 for prototypes). However, the concept of effort in choice has its earlier roots in the concept of decision-making heuristics, examined by Tversky and Kahneman (1974) (for a review, see Kahenman 2004; Kahneman, Slovic, and Tversky 1982; for a review of more recent work on decision-making heuristics, see Todd and Gigerenzer 2000).

A heuristic may be thought of as low-effort quick-and-dirty means of generating what is usually a good enough solution to a decision-making or diagnostic problem in a manner that is more efficient than following a more formal algorithm. The latter approach will usually generate the optimal solution, (i.e., the best outcome possible given the factors at hand) but often at the expense of a great investment of mental effort and time (e.g., the upper limb of Figure 7.1). A good example of the contrast between algorithms and heuristics can be provided by comparing the potential decision strategies for the choice between options. Consider a consumer, choosing among three products or options (e.g., cars) that differ on three attributes (e.g., gas mileage, durability, and price). Typically, the attributes will have different degrees of importance to the consumer—high, medium, and low. Using an optimum decision-making algorithm, the consumer will consider all values of each option on each attribute, weighted by attribute importance. This process, if done mentally, requires an extensive, effortful reliance on working memory operations. Here, the low value of one option on one attribute (e.g., high price) can be compensated for by higher values on the other attributes (e.g., high gas mileage or high vehicle durability). Hence, this method of choice is sometimes described as compensatory (Wickens and Hollands 2000). In contrast, using the heuristic elimination by aspects (EBA) strategy (Tversky 1972), the decider will first eliminate from consideration the options that are least favorable on the most important attribute (e.g., all cars that do not fall within the decision maker's price range) and, hence, will immediately reduce the cognitive working memory load, or effort, of the decision problem.

As an example, the EBA heuristic might quickly narrow a large range of options down to two, leaving two items ranking first and second on the most important aspect. Thus, a car shopper with a hard limit on the amount of money to be spent might quickly reject all models that fall above the price limit and consider only the two cheapest cars from among the available choices. This may not always generate the best solution. One of the eliminated options, for instance, may have actually been better on all of the remaining attributes than were the two options left standing by the EBA strategy. However, the strategy will always lead to an option that is acceptable and will do so with less mental effort than is needed to reach an option that is

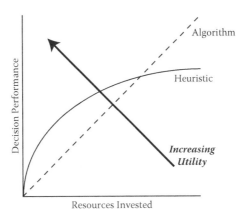

Figure 7.2   The effort–performance representation of algorithms versus heuristics. Note that the higher level of utility of the heuristic explains why it is often chosen over the algorithm.

perfect. The effort-saving benefit of this heuristic will grow, moreover, as the choice space expands to include both more options and more attributes.

Wickens and Hollands (2000) represented algorithms and heuristics in the effort-performance space, as shown in Figure 7.2. Here, algorithms can attain very good performance, but only with maximum investment of effort. Heuristics, on the other hand, can obtain pretty good performance with smaller effort investment. If the utility of a decision strategy can be char-acterized as a weighted sum of performance and effort conservation, then it is easy to think of a utility scale running from the lower right (poor per-formance, high effort) to the upper left (good performance, little effort). This is shown by the arrow in the figure. Within this representation (Navon and Gopher 1979), it is then easy to see how a heuristic can be chosen that has overall greater utility than an algorithm.

In a similar model, Bettman, Johnson, and Payne (1990) presented a con-tingent decision model that describes in detail the role of effort in the choice of decision strategy, identifying six elementary information processing (EIP) mechanisms that are involved to varying degrees in the decision process: READ, ADD, COMPARE, MULTIPLY, DIFFERENCE, ELIMINATE. Their model characterizes different decision strategies based on the number of EIPs they involve. Generally, fewer EIPs cumulate to impose less effort but also tend to reduce decision accuracy. Their model predicts that the choice of a particular strategy will therefore be based on the trade-off between the desired accuracy (i.e., more EIPs needed) and the available effort (i.e., fewer EIPs wanted), corresponding to the top (high accuracy) and bottom (low effort) limbs of Figure 7.1, respectively, as well as to the upper right and lower left regions of Figure 7.2. Their data validate the model.

Bettman, Johnson, and Payne (1990) applied the contingent decision model specifically to consumer choice. As one specific example of such an application, Russo (1977) considered the scenario of consumers faced with a wealth of options at the supermarket.

These options may (1) present separate price and weight information for products ordered or categorized on the shelves by brand; (2) present unit price information (e.g., $/oz) ordered by brand; or (3) present unit price information ordered by unit price value. Analysis can reveal the reduced cognitive effort moving from option 1 to 3 required to select the optimal brand—at least where optimality is defined by dollars per weight—and Russo documented the greater optimality achieved in consumer choice by the third way of presenting information. Bettman, Johnson, and Payne (1990) describe the third procedure of price-ordered marking as an example of *passive decision support*, in which information is conveyed in such a way as to make the least effortful strategies yield the most optimal results. Bettman, Payne, and Staelin (1986) extended this approach to illustrate how information concerning product risks can be presented in a way that demands little effort to access and integrate, here tying back to the issue of safe behavior, which was discussed already. The reader will note the parallel to display layout strategies discussed in chapter 4, where displays were laid out so that expectancy for a pair of displays was inversely correlated with the distance, or effort, between them.

Of course, when the decision maker is confronted with a new way of making choices—or a new kind of choice to be made—he or she may not know the amount of effort that will be required by a particular strategy or the level of accuracy that the strategy can be expected to produce. Thus, the choice will be based on anticipated effort and accuracy. Fennema and Kleinmuntz (1995) examined these anticipated variables and concluded that people are not always highly calibrated in their estimates of these anticipated quantities.

Effort, Salience and Accessibility in Choice Accuracy

The quality of decision making is also influenced by the ease (effort) of processing individual decision cues, conjoined with the salience of those cues (Muthard and Wickens, 2005; Stone Yates and Parker, 1997; Wallsten and Barton, 1982; Wickens and Hollands, 2000). To the extent that one of a set of cues, relevant to a decision, choice, or diagnosis is salient or requires little effort to access and process, this cue will provide a heavy weighting on the outcome; a weighting that is of benefit, if the cue is reliable, and problematic if it is not. The combined influence of salience and effort in this regard can be captured by a construct that Kahneman (2003) has labeled as *accessibility* in an article based on his Nobel Prize address. Accessible information leads to rapid, intuitive diagnosis and choice. Less accessible information may be discounted or underweighted even if it is important.

As an example, the consumer's choice of a multiattribute product may be guided heavily by the manufacturer's strategy to place its most positive attributes in large colorful form at the top of an advertisement (salient and easy to access), while relegating the less attractive attributes at the bottom in fine print. Closely related is the greater salience of the larger ad in the *Yellow Pages*. In information displays, the frequency (count) of objects or events is generally more salient and accessible than the probability of such events, because understanding the latter requires an effort-demanding division by a quantity that is not always salient: the total number of possible events, or the "base rate" (Kaheman, 2003). Hence, people may base decisions more on frequency than probability, even though the latter is often more appropriate and optimal. Finally, in viewing dynamic displays such as those in air traffic control, distance (between an aircraft and another hazard) is generally more salient and accessible than are time or speed, and so the former may dominate judgments of future trajectories at the expense of the latter two, even though these latter two may be more appropriate (Xu and Rantanen, 2007). This may be particularly true if distance is made more salient by enlarging the display (Muthard and Wickens (2005).

Effort in the Choice of a Feature to Use: Implications for System Design

As people work with computers to accomplish tasks, there are often various features available to solve the same problem: Some features may be powerful but demand a heavy investment of effort—either to use (e.g., high working memory load) or to learn (e.g., complex multistep procedures). This describes the upper limb of Figure 7.1. Others may be inefficient but may impose a low cost in effort. An example of the first (upper limb) type of strategy while text editing might be the execution of a few complex and powerful commands in a word processor to rapidly locate a desired word string in a long document. An example of the second (lower limb) type of strategy might be the line-by-line scanning of the text to locate the desired string, a process that is more time consuming but of less cognitive complexity.

We note in this example that effort has a new meaning—not time but cognitive or mental load. In fact, the choice between strategies has at least four important elements: (1) time demands; (2) cognitive demands, which are often closely associated with working memory demands; (3) the accuracy of the output; and (4) the utility of choice, which may reflect a weighted combination of the first three variables. That is, people may be seen to maximize utility by choosing a strategy that achieves some compromise among minimal time, minimal effort, and maximum expected accuracy. Yet circumstances can alter the weighting assigned to each of the three elements, as suggested by three experiments described following.

Ballard, Hayhoe, and Pelz (1995) asked people to copy a pattern of colored blocks—the model—by picking up and transferring blocks from a resource area onto a canvas. In some conditions, the model and the resource area were close together, such that it was possible to make short eye movements back and forth between them. Under these circumstances, Ballard and colleagues found that people tended to minimize the demands on working memory, looking at the model first to determine what color of block was needed next, looking at the resource area to select a block of the appropriate color, and then looking back at the model to determine where the block should be placed. Thus, evidently, no more than one piece of information—color or location—was held in memory at a time, indicating that people preferred many repeated information access actions to fewer actions with higher working memory demands. When the model and resource area were moved farther apart, however, such that more effortful head movements were needed to shift attention between them, performance strategy changed: Participants made fewer gaze shifts between the two areas and appeared to retain both color and location information simultaneously in working memory.

A study by Gray and Fu (2004) likewise observed a preference for subjects to avoid high information-access costs and instead to rely more on memory. Subjects were asked to program a VCR under three conditions: one in which programming information about shows and times was memorized before the experiment, one in which this information was not memorized but could be gained by visual scanning, and one in which the information needed to be accessed by key press requests. Thus, the three conditions could be ordered by ease of information access from easiest (direct memory retrieval) to moderately difficult (visual scanning) to hardest (manual interaction). Data revealed that subjects preferred direct memory retrieval—the learning group did best. Even within the two groups who had to access information perceptually, subjects often relied on imperfect working memory of programming information in lieu of efforts to retrieve information from the display. Moreover, subjects who were required to manually interact with the display to view the programming information relied more on imperfect memory than those requiring less effortful scanning.

Finally, a study by Gray et al. (2006) found that subjects would modulate their dependence on both memory and perceptual mechanisms. Using the similar pattern-copying paradigm to that employed by Ballard, Hayhoe, and Pelz (1995), they observed that as the time cost to access screens was progressively increased by imposing an artificial time delay, thereby discouraging repeated perceptual access, then progressively greater reliance was placed on a working memory strategy. Gray and colleagues describe this as an example of a soft constraint. That is, the human's choice between strategies is based not only on hard physical or information-processing limits of the system (e.g., an action is unavailable) but also is highly contingent on the moment-by-moment information-access costs and memory demands.

Effort in Information Integration: The Proximity Compatibility Principle

A final example of effort in single-task performance relates to the proximity compatibility principle discussed extensively in chapter 6 (Wickens and

Carswell 1995). When information needs to be integrated from two different sources, there is an increased penalty of moving those sources farther apart or making it harder to access one source after attention leaves the other, relative to the case when the two sources are part of separate tasks or separate judgments. This effect is explained in terms of the competition between the effort required to retain information in working memory as attention shifts from one display source to the next and the effort required to access the second source necessary for comparison or integration with the first. These two mental operations—working memory maintenance and information access—which are both effort demanding, compete for the common pool of cognitive resources. The greater the integration requirements (i.e., working memory load) and the greater distance traveled by attention as it moves from the first to access the second source, the greater is the total effort demand of both information processing components and thus the greater is the performance penalty (Vincow and Wickens 1993).

This last example provides an important transition from effort in a single-task context to effort in dual-task performance. The proximity compatibility principle reflects the role of effort in resource competition between different mental operations within a single-integration task. More generally, we now turn to the role of effort in accounting for the failure to divide attention between two tasks—that is, in multitasking or dual-task performance.

Effort as Resources: Dual-Task Performance

Kahneman (1973) presented the most comprehensive theory of mental effort as it is manifest in dual-task performance, equating the effort with the mental resources necessary to sustain both single- and dual-task performance. He posited the limited availability of such mental resources and proposed that when people perform increasingly difficult tasks those resources will be mobilized through increasing arousal of the autonomic nervous system. This arousal has physiological manifestations, and Beatty and Kahneman (1966) and Beatty (1982) have found that such difficulty-induced arousal can be most directly indexed by pupil diameter. In addition to supplying the foundations for Norman and Bobrow's (1975) performance resource function (PRF) (see following), a major component of Kahneman's theory is the view that much of the variance in the amount of mental resources available for performance is explained by the difficulty of the task itself. We might phrase this by saying, "It's hard to try hard on an easy task but easy to try hard on a hard task." Thus, one needs the challenge of a difficult task to fully mobilize resources. However, the resources mobilized with increasing task difficulty are of diminishing availability as the demand increases, thereby yielding a supply–demand curve as illustrated by the curved line in Figure 7.3. As a consequence, the shortfall between resources needed and those supplied (the difference between the straight and curved line) increases as task

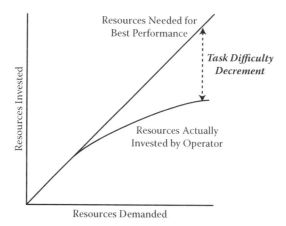

Figure 7.3 The resource supply–demand curve.

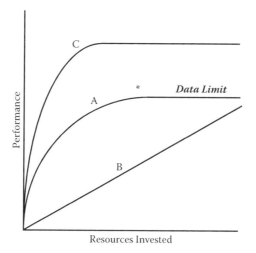

Figure 7.4 Three examples of the PRF. (A) and (C) contain data limits.

demand increases, yielding a progressive loss in performance, which might
be referred to as a task-difficulty decrement.

Norman and Bobrow (1975) carried Kahneman's (1973) qualitative theory
further into the quantitative modeling domain by proposing the PRF and
the performance operating characteristic (POC) trade-off function between
tasks. The PRF, three examples of which are shown in Figure 7.4, characterizes
the performance of individual tasks in either a single- or a dual-task context,
plotting the performance on a given task as a function of the resources invested.
In this regard, the PRF is directly analogous to the function that contrasts
heuristics with algorithms in Figure 7.2. From the PRF emerged the distinction
between resource-limited tasks or regions and data-limited tasks or regions.
The resource-limited tasks are those for which providing more resources

(i.e., trying harder, borrowing resources from a concurrent task) will always improve performance. In Figure 7.2, an algorithm was such a task. The bottom line of Figure 7.4 is a fully resource-limited task (task B). In contrast, data-limited tasks are those in which performance is not limited by the amount of resources supplied (i.e., trying harder and investing more effort will not consistently improve performance) but rather by the quality of data. The middle line (task A) is partially data limited since to the right of the asterisk further investment of resources will lead to no further gains in performance. The top line (task C) represents a task that is almost fully data limited.

Data limits may arise from one of three sources:

(1) There may be poor memory data (e.g., You cannot perform well on a language when you do not know the vocabulary, even if you try hard).

(2) There may be poor perceptual data (e.g., You cannot detect a subthreshold signal no matter how effortfully you strain your eyes).

(3) A task may be data limited if it has obtained full automaticity—as discussed in chapter 2—through high levels of practice with consistently mapped tasks. Here you can perform just about as well as possible while investing few, if any, resources (Figure 7.4, task C curve). It should be noted that examples (1) and (2) have large data limits but low asymptotic levels of performance, like curve A in Figure 7.4.

The PRF can be generated in either of two different ways. One is to induce more or less voluntary effort to be invested into a single-task performance. Vidulich and Wickens (1986) did this through financial incentives, noting improved performance when such incentives are offered. The second is through a sort of reverse engineering from the POC in dual-task performance. The POC (Figure 7.5) is a cross-plot of performance on two time-shared tasks as the priorities are varied between them—that is, by giving instructions that one task is primary and the other is secondary, then reversing the instructions, then giving equal task priority instructions (Navon and Gopher 1979; Sperling and Melchner 1978; Tsang 2006). Such priority trade-off instructions are assumed to implicitly vary the allocation of limited mental resources between the two tasks and, hence, to reflect the dependence of performance of each task on the resources allocated to it (i.e., the PRF). Interestingly, when most PRFs are generated, single-task performance does not lie on the axes where the curve intersects; instead, it is a little better—about where the smiley face is in Figure 7.5a. This difference between doing a task alone and doing it with the possible requirements for another task, even as the latter is greatly deemphasized, is called the *cost of concurrence* (Navon and Gopher 1979) and may be thought of as sort of an executive control overhead associated with managing two tasks, as discussed more fully in Chapter 9.

Though the POC can take a variety of shapes, Figure 7.5 illustrates four different representative categories, with the PRF assumed for the two single tasks shown below. Each of these categories has a qualitatively different

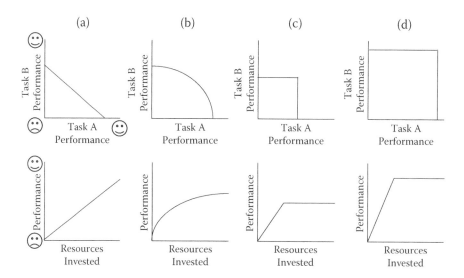

Figure 7.5   For examples of the POC between two tasks A and B (top row), and the corresponding single-task PRF of the two tasks (bottom row). In each example, the two tasks have identical PRFs.

interpretation. Within each POC, single-task performance—when all resources are allocated to the task—is shown where the POC intersects with the axes.

In Figure 7.5a, the linear POC suggests a linear PRF. It might represent performance on a test, where time is the resource given to the test. Every added unit of test-taking time allocated will increase the number of correct answers. Time sharing of taking two tests would be reflected in the POC.

Figure 7.5b illustrates a frequently observed curvilinear POC, generated by a curvilinear PRF as shown below. Here, performance shows a pattern of diminishing returns. Investment of progressively more resources in a task improves performance at a decreasing rate. The similarity of the PRF to that shown in the data-limited range in Figure 7.4 should be noted.

Figure 7.5c shows one of two forms of a box-like POC. Here providing more or fewer resources to the task has little or no impact on performance. Such a pattern is characteristic of the data-limited task described already. However, as shown in Figure 7.5d, sometimes the box-like characteristic of the data-limited PRF exists at perfect performance. It takes few resources to achieve that performance, and allocation of additional resources gains nothing. This mimics the property of curve C in Figure 7.4 and reflects a ceiling effect on highly automated tasks.

The applied importance of the POC and its underlying PRFs is that they can provide guidance as to the optimal allocation of attention in multi-task situations. When the POC involves an entirely linear trade-off as in Figure 7.5a, then this optimal point is simply dictated by the relative costs of doing poorly on each task. If each of the costs of poor performance on the two tasks is equal, there is no optimal. For the other curves, an optimal division of attention can be specified independently of task costs. For example, with two curvilinear PRFs such as shown in Figure 7.5b, the combined best performance on both tasks will be achieved at a 50–50 allocation. If, on the other hand, one task is data limited while the other is resources limited, it makes sense to give more attention to the resource-limited task and only give enough attention to the data-limited task to bring performance to the level of the data limit—that is, stopping when resource allocation reaches the asterisk in Figure 7.4 for task A). Beyond that, additional resources allocated to that task are wasted. Schneider and Fisk (1982) showed that resource-allocation instructions can maximize joint dual-task performance when a data-limited task is time shared with a resource-limited one.

Relevance of Effort Concept for Mental Workload

Underlying the notion that tasks vary in the amount of effort—meaning mental resources they demand or, equivalently, in the amount of mental workload that they impose—is the idea that the effort demand of tasks is an important commodity to be measured (Tsang and Vidulich 2006). We describe three of these circumstances as follows.

Tasks demanding different levels of effort may yield equivalent performance when resources are plentiful, as shown by PRFs A and B in Figure 7.4. However, task A, because it has a greater data-limited region (i.e., is more automated), can avail more spare capacity when it is performed at its maximum level. As a consequence, task A will be less vulnerable to performance loss should unexpected new resource demands be placed on the user. For example, suppose that in comparing two electronic map interfaces for use in a driver navigation system a designer discovers that one interfaces imposes higher demands on working memory than the other, such as by requiring the user to remember and enter map coordinates. Working memory demand is a major source of resource demand. Although the two interfaces may show no difference in supporting navigating performance when performed under single-task conditions, the more memory demanding interface may show greater interference when placed in the dual-task context of driving concurrently with map use, particularly when the driver is confronted with unexpected circumstances that require high resource investment (e.g., encountering an unanticipated detour).

Closely related to this issue is the expansion of the timeline model of mental workload, discussed in the context of single-channel theory in chapter 2. In that context, workload was simply defined as the ratio

$$[\text{time required}]/[\text{time available}].$$

But this simplification may be misleading when tasks that require different amounts of time to perform also differ in the effort they demand during that time. An index of workload should differ depending on whether two constant-time tasks are easy or difficult. Consider the workload of lane keeping while driving on a clear day with a dry road relative to that of driving on a snowy night. In both cases, the time occupied by lane keeping may be the same—an essentially continuous task—but clearly the demands on the snowy night are greater, and its potential to interfere with other tasks will be greater as well. An implication of this is that if there are ways of calculating the resources demanded by a task, the most valid timeline analysis calculation of workload should be based on the ratio

[resources required during time t]/[resources available during time t],

where these terms are the products of time × resource demand. The challenge of how this might be calculated is discussed in the next section.

We have already noted that task-demanded effort, or resource mobilization, is potentially costly to the human system; continued arousal and sustained effort impose an eventual toll. Nowhere is this better illustrated than in the sustained attention or vigilance task (Warm 1984) in which the mental demands of simply watching and waiting for rare events to occur have been well documented (Warm, Denber, and Hancock 1986), even under conditions in which performance does not suffer a great deal. The U.S. Army has considered these fatigue implications of sustained effort costs by mandating different maximum flight times for helicopter pilots depending on the difficulty of the mission. For example, flying at night, particularly with night-vision goggles, is deemed several times as effortful and therefore is given proportionately shorter maximums than flying during the day.

Measuring Effort and Mental Workload

Nearly half a century of work has gone into developing ways of quantifying mental workload, or the effort demand of tasks (e.g., Moray 1979, 1988; Tsang and Vidulich 2006; Williges and Wierwille 1979). In this regard, researchers and workload practitioners have considered at least four different categories of techniques for the assessing mental load of a primary task, a designation we give to the task of interest.

Computed Primary Task Properties

Measuring primary task performance is often necessary but is rarely sufficient to understand the workload imposed by the primary task itself. As the contrast between tasks A and B in Figure 7.4 illustrates, very differently loading primary tasks can yield the same performance. However, many primary tasks can be analyzed in terms of elements that will increase their resource demands. For example, higher working memory demands can almost always be assumed to demand more resources (e.g., the seven-digit phone number versus the phone number with area code), as can higher bandwidth tracking tasks or more complex choice tasks. Thus, even without direct measurement of human performance it is possible to predict the level of resource demand of some of these tasks. Still, a limitation of these techniques is that they do not readily lend themselves to comparisons across qualitatively different tasks (e.g., What level of working memory has equal workload to the resource demand of a 0.5 Hz bandwidth tracking task?).

Secondary Tasks

As noted previously, though primary task performance is not necessarily an adequate index of primary task workload, variations in an operator's performance of a resource-limited secondary task can be used to infer the resource demands of a primary task. The secondary-task approach is based on the simple reasoning that if the primary task requires fewer resources, it will avail more resources for the secondary task, performance of which will increase accordingly. There are scores of different secondary tasks available; examples include mental arithmetic probe RT tasks and time estimation (see, e.g., Tsang and Vidulich 2006; Wickens and Hollands 2000; Williges and Wierwille 1979).

Subjective Measures

There is a long history of having people simply rate their perceived difficulty of performing a task, and a variety of different subjective scales have been developed, such as the Task-Load Index (TLX) (Hart and Staveland 1988), the Subjective Workload Assessment Technique (SWAT) (Reid and Nygren 1988), and the Modified Cooper Harper scale (Wierwille and Casali 1983) (for reviews, see Hill et al. 1992; Wickens and Hollands 2000). Though these ratings are often reliable, they also have their own limitations (Vidulich and Wickens 1986; Yeh and Wickens 1988). Interestingly, the techniques of computational primary task properties and subjective measures are integrated in the now classic McCracken and Aldrich scale of task demands (Aldrich, Szabo, and Bierbaum 1989) in which the authors asked a large panel of workload experts to assign ratings to the difficulty of a set of different component tasks, a sample of which is shown in Table 7.1.

Physiological Measures

Returning to the original Kahneman (1973) model proposing that resource demand is mediated by autonomic nervous system arousal, it is not surprising to note that several investigators have found that some of these measures of autonomic system activity correlate well with other task-demand characteristics. For example, pupil diameter has been found to correlate quite closely with working memory load (Beatty 1982), and several investigators have noted the association between heart-rate variability and task mental

Table 7.1   VACM Values and Descriptors

Value	Visual Scale Descriptor	Value	Auditory Scale Descriptor
0.0	No visual activity	0.0	No auditory activity
1.0	Register/detect image	1.0	Detect/register sound
3.7	Discriminate or detect visual differences	2.0	Orient to sound, general
4.0	Inspect/check (discrete inspection)	4.2	Orient to sound, selective
5.0	Visually locate/align (selective orientation)	4.3	Verify auditory feedback (detect occurrence of anticipated sound)
5.4	Visually track/follow (maintain orientation)	4.9	Interpret semantic content (speech)
5.9	Visually read (symbol)	6.6	Discriminate sound characteristics
7.0	Visually scan/search/monitor (continuous/serial inspection, multiple conditions)	7.0	Interpret sound patterns (e.g., pulse rates)

Value	Cognitive Scale Descriptor	Value	Motor Scale Descriptor
0.0	No cognitive activity	0.0	No motor activity
1.0	Automatic (simple association)	1.0	Speech
1.2	Alternative selection	2.2	Discrete actuation (button, toggle, trigger)
3.7	Sign/signal recognition	2.6	Continuous adjusting (flight control, sensor control)
4.6	Evaluation/judgment (consider single aspect)	4.6	Manipulative
5.3	Encoding/decoding, recall	5.8	Discrete adjusting (rotary, vertical thumbwheel, lever position)
6.8	Evaluation/judgment (consider several aspects)	6.5	Symbolic production (writing)
7.0	Estimation, calculation, conversion	7.0	Serial discrete manipulation (keyboard entries)

Notes: Example of the McCracken and Aldrich (1984) scale. Within each of four categories, different tasks are assigned different levels of difficulty or resource demand.

load (e.g., Mulder and Mulder 1981; Sirevaag et al. 1993; Vicente, Thornton, and Moray 1987). Kramer and Weber (2000) reviewed many of these techniques, and chapter 11 in this volume discusses others related to brain activity.

Conclusions to Workload Measurement

Each of the aforementioned techniques has its strengths and weaknesses, and it is by now well accepted that the best workload assessment methods involve combinations of tools and avoid certain tool applications in circumstances that highlight the weaknesses (e.g., using pupil diameter to assess mental load in environments with rapidly changing illumination or heart-rate variability in tasks that also have a high physical workload demand). It is important to note, however, that the techniques of computational task properties and table look-ups are the only ones in which workload demands can be inferred in the absence of actually measuring human performance or physiology. Hence, these techniques are of the greatest value for computational modeling of human performance and workload (Laughery, LeBiere, and Archer 2006).

Effort Investment in Learning, Retention, and Education

The previous discussion has focused on the effort demand of tasks. This final section returns to consideration of the effort invested into a particular mental operation in a learning environment. The role of effort here is the obvious one in which investing more effort into the learning process will lead to longer-term retention, or memory, of the studied or practiced material. This commonsense relationship in fact has several important implications for instruction and design, not all of which are intuitively obvious (Paas, Renkl, and Sweller 2003). The following are four examples.

Deep versus Shallow Processing

Craik and Lockhart (1972) have well documented the advantage of a deep-processing strategy for encoding new information into long-term memory, in which effortful attempts are made to relate the new material to material that is already in long-term memory (ponder its meaning) or to form mental images (Leahy and Sweller 2004, 2005); this is in contrast to a shallow-processing strategy of simply engaging in rote rehearsal—circulating the sound of the new material in verbal short-term memory—which is often a relatively effort-free activity.

The Generation Effect

The *generation effect* describes the general principle that when an action is self-generated, the action itself—and the circumstances surrounding that action—will be better retained than if the same action were performed passively or were witnessed as another agent performed it (Richland, Linn, and Bjork 2007; Slameka and Graf 1978). For example, we might apply this to the better memory we will acquire for learning a traveled route if we actually drove the route than if we were a passenger in a vehicle driven by another (Gale, Golledge, and Pellegrino 1990; Williams, Wickens, and Hutchinson 1996). The significance here is that greater effort is required to generate the action than to passively watch it, and this effort is productively applied to learning about the domain within which the effortful actions were generated.

This concept has been applied to problems that pilots have in understanding the state of some of the complex automation systems they have in the cockpit (Sarter and Woods 1997). As discussed in chapter 3, although there are display annunciators that inform the pilot what mode of operation the plane is programmed to be in, analysis of both accident and incident reports indicate that pilots are often not adequately aware of the current mode: low situation awareness. Put simply, because the pilots themselves did not actively fly or sometimes even choose these modes, their memory, or knowledge, of them is degraded compared with circumstances in which the active mode choices would be actively made by the pilot. More generally, automation designers should strive to keep users sufficiently in the loop so that the effort invested in system control and supervision will avail good knowledge of the current state (Endsley and Kiris 1995; Wickens, Lee, et al. 2004). One simple way of doing this is to periodically disengage the automation-and-return control of the system to the operator (Parasuraman, Mouloua, and Molloy 1996), giving him or her additional incentive to remain aware of system status. Such a manipulation has been shown to decrease the risk of out-of-the-loop behavior, even after control is returned from the human operator to the automation.

Active Learning

The generation effect has its parallels in the frequently advocated and well-documented advantage for active learning in many classroom environments (Mayer 1999), in which learners attempt to solve problems or produce knowledge rather than being passively presented with information. As with generating actions, the process of generating problem solutions is demanding, but the effort is generally productive in forming the necessary long-term memories—or updating short-term situation awareness—to support better retrieval. An alternative form of active learning—one that may be useful in cases where the material under study is too complex for the student to self-generate—is self-explanation. Like solution generation, self-explanation encourages the learner to effortfully relate the material under study to existing information in long-term memory, thus improving the acquisition and retention of new knowledge (Chi 2000).

Learning Performance Dissociation Mediated by Effort

In a review of the skill acquisition literature, Schmidt and Bjork (1992) noted the counterintuitive finding that training techniques producing rapid skill acquisition may actually inhibit long-term retention and transfer of the trained skill. Often, the relative quality of skill learning versus skill retention and transfer hinges on the level of effort demanded during learning.

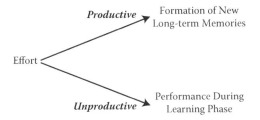

Figure 7.6   Productive and nonproductive effort in learning.

More specifically, data suggest that retention and generalization of skills and knowledge are often degraded when the learning environment is designed to reduce effort. One example is error prevention. Error-prevention learning techniques, like training wheels on a bike (Carroll and Carrithers 1984), during learning leads to both effective performance and relatively low effort investment during the learning process, since the learner often does not need to think much about what he or she is doing but just follows the guidance. But the consequence of the failure to invest effort—to think about what is being done—is that more general principles of the learning environment are not mastered.

In contrast to the aforementioned examples, which encourage effort investment, however, there are circumstances in which effort investment in the learning process may actually be counterproductive. These circumstances are revealed in part by the representation in Figure 7.6, which identifies the two destinations of effort investment: (1) productive effort, which is the formation of new memories (i.e., discovering the skills and consistencies in the environment to be learned); and (2) nonproductive effort, which is performance of the skills required in the learning environment unrelated to the concepts to be learned. If too much effort is allocated to the latter, it may actually inhibit the former (Paas, Renkl, and Sweller 2003). Two examples illustrate.

Sweller (1994) examined the role of problem solving (active learning) in mastering concepts and noted that if these problems are too difficult or the learner is left to mentally thrash about, then so much effort may be invested in just trying to solve the problem that the higher-level principles illustrated by the problem will be missed. For this reason he advocated the use of worked examples in the problem set that can gradually lead the user through the problem on more or less the right track (Sweller and Cooper 1985). This has a more general application to uncontrolled versus error-free guided training discussed already. Although uncontrolled training can sometimes lead to more productive effort investment, as noted, it also has the potential drawback of inducing repetition of errors and mental thrashing. Although the commitment of one error in learning may be quite advantageous (i.e., the learner learns how to correct it) excessive error repetition may be counterproductive (i.e., the learner undesirably learning the erroneous action through repetition).

There is always a danger that introducing new educational technology with glitzy features designed to interest and motivate the student (e.g., invest effort) may lead to investment of nonproductive effort into those game-like features unrelated to the material to be learned. Some aspects of virtual reality technology fall into this category (Wickens 1992; Wickens and Baker 1995). Animation also has been associated with such nonproductive effort, sometimes making multimedia instruction highly interesting but ironically bypassing some of the productive effort investment necessary to understand concepts when they are somewhat less intuitively portrayed (Mayer et al. 2005).

In summary, there is no doubt that effort investment is necessary to efficiently form new memories during learning or training. Such investment can be productive for learning, even if this may yield performance challenges and difficulties during training. But it may also be counterproductive if resources are diverted from productive purposes to either difficult or interesting but irrelevant aspects of the learning environment. The challenge in training system design is to make this careful distinction and to assess how the combined factors of the complexity of the material and the knowledge possessed by the student invite the appropriate level of effort investment into pedagogical techniques. The cognitive load theory nicely integrates this perspective on effort and learning (Mayer 1999; Mayer and Moreno 2003; Paas, Renki, and Sweller 2003; Sweller 1994).

Illusory Competence

Finally, the concept of effort is integral to an intriguing learning-related phenomenon that Bjork (1999) described as the illusion of competence in our own memory capabilities, a phenomenon that has considerable applications for learning and training. In short, people often think that they have learned things better than they actually have. Bjork's analysis has two components. First, as already noted, characteristics of a task that demand productive effort investment in learning may inhibit performance during learning or training but will increase memory and therefore retention of that material. Second, people intuitively evaluate the ease of learning, training, and practice as a proxy for the quality and effectiveness of that learning: People erroneously think that if learning is easy it is effective and that memory for what is learned will be strong. This is an illusion. People using this heuristic—ease of learning = quality of learning—will often study material less than they should, or choose an inappropriate easy training technique (e.g., relying on training wheels).

Conclusions

In conclusion, we have seen the importance of the concept of mental effort or mental (attentional) resources in both single-task performance, which accounts for the choice of behavior and the quality of learning, and dual-task performance, which accounts for the interference between tasks. A critical challenge for applied psychologists remains to be able to quantitatively measure these constructs of effort, difficulty or resource demands, and the domain of mental workload measurement provides some of the material needed to address this challenge.

The resource demand, or difficulty, of a task is intuitive and is a strong predictor in accounting for differences in the success of divided attention, or time sharing, between tasks. Such time sharing is easier and dual-task performance is better with easier tasks demanding fewer resources than with harder tasks demanding more. However, it turns out that resource demand is only one of four important factors that account for the success of performance in multitask environments. The second of these, discussed already, is the resource-allocation policy: Favored and emphasized primary tasks will be performed better than unfavored secondary tasks. The other two factors are emergent properties of the dual-task set, their demand for multiple resources, and the similarity of their information-processing routines. The next chapter turns to these issues and presents a general model of time sharing.

8

Time Sharing and Multiple Resource Theory

Introduction

Driving along a crowded highway on a rainy evening while trying to glance at the map and search the roadside for the correct exit, the driver's cellular phone suddenly rings. The driver feels compelled to answer it and to engage in the conversation with the caller. Will the driver be successful in this multi-task endeavor? What is the likelihood that this added demand will seriously impair safety? Could a different interface on the phone make a difference? Suppose the map was displayed in a head-up location. Would the benefits of not having to look downward be offset by the clutter costs of trying to see two overlapping images? This chapter presents a general model of time sharing that can provide the basis for answering such questions.

The previous chapter presented a model of resource demand and allocation that implicitly assumed that these resources were unitary—a general pool of mental effort—and that it did not matter much whether tasks were visual, auditory, spatial, linguistic, perceptual, or action oriented. In that undifferentiated resource model, the key feature in predicting time-sharing interference was the demand for resources and the resource-allocation policy. Yet some obvious observations inform us that other factors are at work in dictating time-sharing efficiency. As one obvious example, it is harder—and thus more dangerous—to drive while reading a book than while listening to the same book on tape. Here the time-sharing efficiency of the two activities is greatly improved by using auditory rather than visual input channels for language processing. In the example in the first paragraph, it is unlikely that the single-task demand of reading navigational information from a head-up display versus a head-down map would differ much, but the difference in dual-task interference with driving might be considerable. This chapter considers four important factors that determine interference differences above and beyond those attributable to single-task difficulty: (1) multiple resources; (2) preemption; (3) similarity-induced confusion; and (4) similarity-induced cooperation.

Multiple Resource Theory

Multiple resource theory is a theory of divided attention between tasks—typical of those carried out by the driver in the opening example—which has both practical and theoretical implications (Wickens 1984, 1991, 2002). The practical implications stem from the predictions the theory makes regarding the human operator's ability to perform in high-workload, multitask environments, such as the automobile in heavy traffic, the aircraft cockpit during landing, or the front office of a business during peak work hours. These practical implications are often expressed in a particular instantiation of multiple resource theory, which we identify as a multiple resource model (Wickens 2006). In the applied context, the value of such models lies in their ability to predict operationally meaningful differences in performance in a multitask setting that result from changes—in the operator or in the task design—that can be easily coded by the analyst and the designer (e.g., Should we use a joystick or voice control in a multitask setting?).

In the theoretical context, the importance of the multiple resource concept lies in its ability to predict differences in dual-task interference levels between concurrently performed tasks, differences that are consistent with the neurophysiological mechanisms: dichotomies defining separate resources as these underlie task performance. The goal of the theory is to account for variability in task interference that cannot easily be explained by simpler models of human information processing such as the bottleneck and filter theories discussed in chapter 2.

In both applied and theoretical contexts, the distinction between *multiple* and *resources* is critical, and this distinction will remain an important theme throughout this chapter. The concept of *resources* discussed in the previous chapter connotes something that is both limited and allocatable; that is, it can be distributed between tasks (Tsang 2006). The concept of *multiple* connotes parallel, separate, or relatively independent processing, as characteristic, for example, of Treisman's (1986) perceptual analyzers discussed in chapter 6. Multiple resources formally concern the intersection between these two concepts, but each concept on its own has much to contribute to an understanding of time-sharing, or multiple-task, performance.

The remainder of the chapter first traces the origins and tenets of multiple resource theory and then describes one particular version of the theory: the four-dimensional model proposed by Wickens (1980) and elaborated on by Wickens and Hollands (2000) and Wickens (2002, 2005a). We demonstrate how this model can be implemented in a computational form and conclude by describing three alternative mechanisms to account for differences in divided attention and dual-task performance.

History and Origins

The origins of multiple resource theory can be traced originally to the concept of a single-channel bottleneck in human information processing, a bottleneck that limited the ability to perform two high-speed tasks together as effectively as either could be performed alone (Broadbent 1958; Craik 1947; Welford 1967). As discussed in chapter 2, such a view was very prominent in the analysis of high-speed tasks (i.e., reaction time tasks in the psychologist's laboratory) and suggested that time was a very limited resource that could not be shared between tasks. As discussed in the previous chapter, however, the concept of effort as a continuously allocatable and sharable resource emerged later and proved particularly attractive in the context of mental workload measurement (Moray 1967). Again, however, the concept was unable to fully account for performance prediction within a multitask context. Thus, it was elaborated on to incorporate multiple resources.

Subsequent to the development of a general resource model of task interference (Kahneman 1973), evidence emerged that considerable variance in dual-task performance could not be attributed just to the difficulty (i.e., quantitative resource demand) of one or more component tasks or to the resource-allocation policy between them (i.e., which task is favored and which is neglected). Instead, evidence was provided that differences in the qualitative demands for information-processing structures led to differences in time-sharing efficiency (e.g., Kantowitz and Knight 1976; Wickens 1976). Such structures thus behaved as if they were separate, or limited, resources. Time sharing between two tasks was more efficient if the two used different structures than if they used common structures (ibid.).

An obvious example of such a structural distinction is between the eyes (i.e., visual processing) and the ears (i.e., auditory processing). In many circumstances dual-task performance is poorer when two visual tasks must be time shared than in a configuration in which the equivalent information for one of the tasks is presented auditorally (e.g., Treisman and Davies 1973). To cite a more concrete example, the vehicle driver will have more success at driving and comprehension while listening to a set of instructions than while reading the same set (Parkes and Coleman 1990). That is, the eyes and ears behave as if they are supported by separate resources. Wickens (1980) performed a sort of meta-analysis of a wide variety of multiple-task experiments in which structural changes between task pairs had been compared and found strong evidence that certain structural dichotomies (e.g., auditory versus visual processing), described in more detail following, behaved like separate resources.

It should be noted here that this aspect of multiplicity—to make parallel processing more feasible and to improve the level of multiple-task performance—does not necessarily have to be linked to a resource concept. For example, chapter 6 discusses multiplicity of perceptual analyzers proposed by Treisman (1986) as supporting parallel processing. However, in a classic article, Navon and Gopher (1979) laid out the clear intersection between the multiplicity and the resource, or demand-level, components in the context of the economics theory of scarce resources. Their theory made explicit predictions about the different trade-offs between two time-shared tasks as a function of their degree of shared resources, their quantitative resource demands, and the resource allocation policy adopted by the performer. In a parallel effort, as noted already, Wickens (1980) then identified the particular

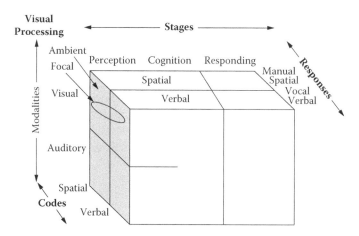

Figure 8.1   The structure of multiple resources (after Wickens 2002).

structural dimensions of human information processing that met the joint criteria of accounting for changes in time-sharing efficiency and of being associated with neurophysiological mechanisms that might define resources. This particular set of dimensions provided the basis for the particular multiple resource model, which we now describe.

The Four-Dimensional Multiple Resource Model

The multiple resource model (Wickens 1984, 1991, 2002, 2005a) proposes that four important categorical and dichotomous dimensions account for variance in time-sharing performance. That is, each dimension has two discrete levels. All other things being equal—equal resource demand or single-task difficulty—two tasks that both demand one level of a given dimension (e.g., two tasks both demanding visual perception) will interfere with each other more than two tasks that demand separate levels on the dimension (e.g., one visual and one auditory task). The four dimensions, shown schematically in Figure 8.1 and described in greater detail, are (1) processing stages, (2) perceptual modalities, (3) visual channels, and (4) processing codes. Consistent with the theoretical context of multiple resources, all of these dichotomies can be associated with distinct physiological mechanisms.

Stages

The resources used for perceptual activities and for cognitive activities (e.g., involving working memory) appear to be the same and are functionally separate from those underlying the selection and execution of responses (Figure 8.2). Evidence for this dichotomy is provided when the difficulty of responding in one task is varied—that is, it demands greater or fewer resources—and this manipulation does not affect performance of

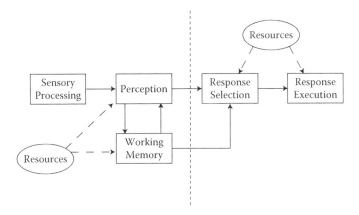

Figure 8.2   Illustrates stage-defined resources.

a concurrent task in which the demands are more perceptual and cognitive
in nature or, conversely, when increases in perceptual-cognitive difficulty
do not much influence the performance of a concurrent task in which the
demands are primarily response related (Wickens and Kessel 1980; Wickens
1980). In the realm of language, Shallice, McLeod, and Lewis (1985) examined
dual-task performance on a series of tasks involving speech recognition (i.e.,
perception) and production (i.e., response) and concluded that the resources
underlying these two processes are somewhat separate, even as they share
linguistic resources. It is important that the stage dichotomy can be associated with different brain structures; that is, speech and motor activity tend
to be controlled by frontal regions in the brain—the forward of the central
sulcus—whereas perceptual and language comprehension activity tends to
be posterior of the central sulcus. Physiological support for the dichotomy
is also provided by research on event-related brain potentials (e.g., Isreal,
Chesney, et al. 1980; also see chapter 11 in this volume).

As an operational example of separate stage-defined resources, the stage
dichotomy would predict that the added requirement for an air-traffic
controller to acknowledge vocally or manually (i.e., a response demand)
each change in aircraft state would not greatly disrupt his or her ability
to maintain an accurate mental model of the airspace (i.e., a perceptual-
cognitive demand).

As shown in Figure 8.2, the stage dichotomy of the multiple resource
model also predicts that there will be substantial interference between
resource-demanding perceptual tasks and cognitive tasks involving working memory to store or transform information (Liu and Wickens 1992a). Even
though these do constitute different stages of information processing, they
are supported by common resources. For example, visual search coupled
with mental rotation or speech comprehension coupled with verbal rehearsal
both provide examples of operations at different stages—perceptual and
cognitive—that will still compete for common stage-defined resources and

will thus be likely to interfere. The cognitive processes in cell-phone conversation, for example, clearly interfere with perceptual processes involved in noting changes in the driving environment (McCarley, Vais et al. 2004).

Finally, we note how the stage dichotomy of multiple resources is consistent with the evidence for a bottleneck in response selection, as discussed in chapter 2 (Pashler 1998), in that two tasks both involving a response-selection stage will heavily compete for the common response-related resource, thus causing a delay in the second-arriving stimulus, but such response selection will compete far less with tasks that rely on perceptual-cognitive processing.

Perceptual Modalities

It is apparent that we can sometimes divide attention between the eye and ear better than between two auditory channels or two visual channels. That is, cross-modal time sharing is better than intramodal time sharing. As examples, Wickens, Sandry, and Vidulich (1983) found advantages to cross-modal over intramodal displays in both a laboratory tracking experiment and in a fairly complex flight simulation, and Wickens et al. (2003) solidly replicated these results. Parkes and Coleman (1990) and Srinivasin and Jovanis (1997) found that in-vehicle information was better presented auditorily, via speech, than visually, via map, while subjects were concurrently driving a simulated vehicle—driving has heavy visual attention demands. Wickens (1980) reviewed several other studies that report similar cross-modal advantages.

The relative advantage of cross-modal, or auditory-visual (AV) over intramodal visual–visual and auditory–auditory (VV and AA) time sharing may not, however, be the result of separate perceptual resources within the brain but rather the result of the peripheral factors that place the two intramodal conditions, VV and AA, at a disadvantage. Thus, two competing visual channels, VV, if they are far enough apart, will require visual scanning between them, which is an added cost. If they are too close together they may impose confusion and masking, just as two auditory messages, AA, may mask one another if they occupy nearby or overlapping temporal frequencies, or pitch. This is sometimes referred to as structural interference.

The degree to which peripheral rather than or in addition to central factors are responsible for the examples of better cross-modal time sharing—AV better than AA or VV—remains uncertain, and when visual scanning is carefully controlled cross-modal displays do not always produce better time sharing (Horrey and Wickens 2004a; Wickens and Colcombe 2007b; Wickens, Dixon, and Seppelt 2005; Wickens and Liu 1988). However, in most real-world settings visual scanning is enough of a penalty for VV interfaces that dual-task interference can be reduced by off-loading some information channels from the visual to the auditory modality in environments such as the anesthesiology workstation (Watson and Sanderson 2004) or the airplane cockpit (Wickens et al. 2003). Furthermore, simultaneous auditory messages

(AA) are sufficiently hard to process so that an advantage can usually be gained by displaying one of them visually (Rollins and Hendricks 1980).

Thus, the issue of whether the advantage of separating auditory and visual displays is entirely a structural or sensory phenomenon related to visual scanning and auditory masking in the intramodality case or whether there are separate auditory and visual resources within perception is one that remains unresolved. It is clear that some experiments that have carefully controlled these peripheral sensory factors have found cross-modal advantages (for a review see Wickens 1980, 1984, 1991). However, it is equally clear that some nonresource factors may offset a separate-resource advantage, in particular the attention-capture, or preemptive, characteristics of auditory information, an issue discussed in greater detail later in this chapter (Spence and Driver 1997; Wickens and Liu 1988).

Finally, although it is not represented in Figure 8.1, there is emerging evidence that tactile information may behave somewhat like auditory channels in terms of its degree of conflict with other resources.

Visual Channels

In addition to the distinction between auditory and visual modalities of processing, there is good evidence that two aspects of visual processing—focal and ambient vision—constitute separate resources in the sense of (1) supporting efficient time sharing; (2) being characterized by qualitatively different brain structures; and (3) being associated with qualitatively different types of information processing (Horrey, Wickens, and Consalus 2006; Leibowitz et al. 1982; Previc 1998, 2000; Weinstein and Wickens 1992; Wickens and Horrey in press). Focal vision, which is nearly always foveal, is required for perceiving fine detail and pattern and object recognition (e.g., reading text, identifying small objects). In contrast, ambient vision heavily, but not exclusively, involves peripheral vision and is used for perceiving orientation and ego motion—the direction and speed with which one moves through the environment. When we manage to successfully walk down a corridor while reading a book, we are exploiting the parallel processing or capabilities of focal and ambient vision, just as we are when keeping the car moving forward in the center of the lane (i.e., ambient vision) while reading a road sign, glancing at the rearview mirror, or recognizing a hazardous object in the middle of the road (i.e., focal vision) (Horrey, Wickens, and Consalus 2006). Aircraft designers have considered several ways of exploiting ambient vision to provide guidance and alerting information to pilots while their focal vision is heavily loaded by perceiving specific channels of displayed instrument information (Nikolic and Sarter 2001; Reising et al. 1998; Stokes, Wickens, and Kite 1990).

Processing Codes

This dimension reflects the distinction between analog–spatial processing and categorical–symbolic (usually linguistic or verbal) processing. Data from multiple-task studies (Wickens 1980) indicate that spatial and verbal processes, or codes—whether functioning in perception, cognition, or response (i.e., all three stages of processing in Figure 8.1)—depend on separate resources and that this separation can often be associated with the two cerebral hemispheres (Polson and Friedman 1988). For parallel views on the important distinctions between spatial and verbal working memory or cognitive operations see also Baddeley (1986) and Logie (1995).

The distinction between spatial and verbal resources accounts for the relatively high degree of efficiency with which manual and vocal responses can be time shared, assuming that manual responses are usually spatial in nature (e.g., tracking, steering, joystick, or mouse movement) and that vocal ones are usually verbal. In this regard several investigations (e.g., Martin 1989; Sarno and Wickens 1995; Tsang and Wickens 1988; Vidulich 1988; Wickens and Liu 1988; Wickens, Sandry, and Vidulich 1983) have shown that continuous manual tracking and a discrete verbal task are time shared more efficiently when the discrete task employs vocal as opposed to manual responses. Also consistent is the finding that discrete manual responses using the nontracking hand appear to interrupt the continuous flow of the manual tracking response, whereas discrete vocal responses do not (Wickens and Liu 1988). Note that a hybrid operation is keyboarding, or typing. This can best be described as a manual response that is fed by verbal cognition—or visual-verbal input, if it is simple transcription—in the context of Figure 8.1.

An important practical implication of the processing codes distinction is the ability to predict when it might or might not be advantageous to employ vocal versus manual control. Manual control may disrupt performance in a task environment imposing demands on spatial working memory (e.g., driving), whereas voice control may disrupt performance of tasks with heavy verbal demands or may be disrupted by those tasks depending on resource-allocation policy. Thus, for example, the model predicts the potential dangers of manual dialing of cellular phones given the visual, spatial, and manual demands of vehicle driving, and it suggests the considerable benefits to be gained from voice dialing (Dingus et al. 2006; Goodman et al. 1999). Note that this benefit clearly does not imply that hands-free phones will be interference free; there will be plenty of resource competition between conversing and driving at the perceptual-cognitive stage (Strayer and Drews 2007). The model also predicts, and research confirms, the greater facility, and reduced interference, of using voice control rather than touch-screen or keyboard control for other in-vehicle tasks while driving (Ranney, Harbluk, and Noy 2005; Tsimhoni, Smith, and Green 2004). Some component of this high-manual interference may be related to the visual resource competition between driving and processing the visuospatial information required to enter the keyboard input.

The verbal-spatial code dichotomy also accounts for the greater disruption of background music when it has words, or lyrics, than when it does not in the typical office environment in which verbal processing is heavily employed (Martin, Wogalter, and Forlano 1988). In driving, it can account for the greater interference between driving and navigating with a memorized map (i.e., spatial) than between driving and a memorized route list (i.e., verbal (Wetherell 1979).

The Multiple Resource Model Revisited

Figure 8.1 presents the four dimensions of the model in a graphical form. Each line boundary in the three-dimensional cube separates the two categorical levels of each dimension (i.e., separate resources). The figure shows that the distinction between verbal and spatial codes is preserved across all stages of processing and that the stage-defined resources are preserved in both verbal and spatial processing. It also depicts the way in which the distinction between auditory and visual processing is defined at perception, but not within cognitive processing, and the way in which the distinction between ambient and focal vision is nested only within the visual resources. Thus, within this dichotomous dimensional structure, to the extent that two tasks share more common levels along more of the four dimensions there will be greater interference between them.

Another important issue that cannot be neglected is that of compatibility. When the designer makes a structural change to a task (e.g., a change from a visual to an auditory display to reduce VV interference), it may be that the new interface is less compatible with the central processing demands of the task. For example, visual arrows are a more compatible way of signaling direction and magnitude than are words. The performance gains purchased by reducing resource conflict can therefore be offset by the increased resource demands produced by a less compatible interface. This issue of stimulus-central processing-response (SCR) compatibility and its effects on resource demand are discussed by Vidulich and Wickens (1985), Wickens, Sandry, and Vidulich (1983), and Wickens, Vidulich, and Sandry-Garza (1984).

To correct a common misconception, it is important to note that the multiple resource model does not predict perfect parallel processing whenever two separate resources are used for two different tasks. This should be evident from the model structure in Figure 8.1. For example, just because auditory and visual tasks use separate resources on the modality dimension, they may still share demand for the common perceptual-cognitive resources within the stage dichotomy (e.g., consider reading while listening to a conversation).

In considering the previous issue, it is also important to raise the question of whether there is any general pool of resources for which all tasks compete (e.g., accounting for interference between tasks that share no common resources). For example, Strayer and Johnston (2001) argued that interference between cell-phone use and driving is a consequence of competition for general attention rather than for specific multiple resources. On the one hand, this issue is to some extent beyond the scope of multiple resource theory, since the theory and model are designed to predict differences in interference as changes in resource competition are imposed—rather than the presence or absence of any interference at all.

On the other hand, this issue can be best addressed by focusing on the allocation component of multiple resources. It is clear that driving and cell-phone conversation compete for some common resources—hence their occasional interference; the precise composition of these resources remains to be fully determined. Much of the time, driving is minimally disrupted by cell phone conversation, but on the occasions where serious violations of driving safety do occur it is obvious, almost by definition, that drivers have failed to prioritize the two tasks appropriately, allocating more resources to the conversation than is warranted. This issue of resource allocation directly invokes the concept of executive control of attention, a concept that is discussed in the next chapter.

A Computational Model of Multiple Resource Interference

As represented in Figure 8.1, multiple resource theory makes qualitative predictions about multiple-task performance by implying that two tasks sharing common resources (i.e., sharing a common level on a dimension) will interfere more than two using separate resources. Recently, efforts have been made to extend such qualitative predictions to more quantitative ones, whereby some measure of absolute predicted performance—or dual-task performance decrements—can be derived to compare a set of systems, tasks, or interfaces (Horrey and Wickens 2003; Sarno and Wickens 1995; Wickens 2002, 2005a; Wickens, Dixon, and Ambinder 2006). Such models typically include two additive components: one based on the total demand (effort component), and one based on multiple resource interference.

The total demand component reflects the plenary demands of the concurrent tasks and can be estimated by simply assigning each task a value of 0 (fully automated), 1 (easy), or 2 (difficult) and summing the values across tasks, thus availing a predicted range of scores between 0 (minimum) and 4 (maximum) in a dual-task situation. More elaborate scales, such as that depicted in Table 7.1 from the previous chapter, can be used (Laughery, LeBiere, and Archer 2006). The resource-sharing component may be computed by establishing the number of the four dimensions along which the two tasks share common resources, here again a value that could range between 0 and 4. In a simplified version of such a model, these two components—added together and equally weighted—provide a predicted total interference measure that could range between 0 and 8.

Such a model has been validated to account for a good deal of variance in dual-task interference—more than 50 percent—across a set of heterogeneous task combinations and interfaces used in aviation (Sarno and Wickens 1995), driving (Horrey and Wickens 2003), and robotics and unmanned air vehicle control (Wickens, Dixon, and Ambinder 2006).

Importantly, though the model predicts the total interference between two tasks, it says nothing about the extent to which one task or the other bears the brunt of the interference—that is, the resource-allocation policy. For example, a greater potential for interference between driving and manual cell-phone dialing versus voice dialing might be of little consequence to safety if the only effect was to hinder the dialing itself. However, if the added interference is partially or fully manifest in driving performance, then the difference would have serious safety consequences. Thus, the multiple resource prediction must be accompanied by a prediction of the allocation or attention management policy of resources between the competing tasks, just as was the case with the single-resource model, as Figure 7.4 in the previous chapter showed. The modeling of the joint effects of priority and multiple resources is complex (Navon and Gopher 1979; Tsang 2006; Tsang and Wickens 1988; Salvucci and Taatgen 2008) and remains in need of validation. Such validation invites the possibility that issues of predicting selective attention to channels and task, captured by the SEEV model (chapter 4 in this volume), can be linked to those of allocation, embedded within the multiple resource model (Wickens 2007). Further discussion of resource allocation policy in the context of executive control and task management is covered in the next chapter.

Other Sources of Interference: Preemption, Confusion, and Cooperation

A hallmark of the mechanisms of dual-task interference that we have described so far is that they are associated with resource demand, strategic allocation, and four relatively gross and anatomically defined dichotomies within the brain. Yet there appear to be additional sources of variance in time-sharing efficiency that cannot be described by these mechanisms but, instead, are related to other characteristics of human information processing. We describe here briefly the role of preemption and, in more detail, the roles of cooperation and confusion.

Auditory Preemption

As noted already, the research findings on whether information delivery along separate perceptual modalities (AV) or shared modalities (VV) supports better dual-task performance remains ambiguous. One reason for such ambiguity is that delivery of a discrete auditory message in the context of a concurrent visual task will be more likely to capture attention and to preempt performance of that ongoing task than will delivery of the same message visually (Helleberg and Wickens 2003; Horrey and Wickens 2004; Iani and Wickens 2007; Latorella 1996; Spence and Driver 1997; Wickens and Colcombe 2007; Wickens, Dixon, and Seppelt 2005). This asymmetry has a number of underlying causes, including the need to rehearse a long auditory message as soon as it is heard—an operation not necessary for a printed visual message (Helleberg and Wickens 2003)—and the intrinsic alerting properties of the auditory channel, as discussed in chapter 3 and in the next chapter. Importantly, the preemption mechanism makes the identical predictions to multiple resource theory when performance of the task whose modality is varied is considered—auditory delivery is better than visual. However, opposite predictions of the two theories are offered when considering performance of the concurrent, usually visual, task: multiple resource theory favoring auditory delivery and preemption theory favoring visual delivery.

In resolving these competing explanations for modality interference, two points emerge. First, the two theories or mechanisms are not incompatible or mutually exclusive. They can both function at the same time and can offset each other, such that there may be little difference at all in performance of an ongoing visual task as a function of whether the interrupting task is auditory or visual. For example, Horrey and Wickens (2004a) found essentially equivalent interference with driving between an auditory and visual (HUD) presentation of in-vehicle task information. Second, the advantage of cross-modal (AV) over intermodal (VV) performance for both tasks grows as the separation between visual sources in the latter condition increases, extending into the eye and head fields and thus increasing the need for visual scanning, as discussed in chapter 4 (Wickens, Dixon, and Seppelt 2002).

Cooperation

The improvement of time-sharing efficiency by increasing similarity results from circumstances in which a common display property, mental set, processing routine, or timing mechanism can be cooperatively shared in the service of two tasks that are performed concurrently. Chapter 6 noted how the close proximity fostered by a single object can improve parallel perceptual processing (e.g., Duncan 1984). Such object-based proximity, as well as other attributes of similarity between two display sources, has been found to improve performance of concurrent tracking tasks (Fracker and Wickens 1989), such as the lateral and vertical dimensions of aircraft control (Haskell and Wickens 1993) or of tracking and discrete tasks (Kramer, Wickens, and Donchin 1985). That is, integrating information for two tasks into a common object allows cooperative perceptual processing of the two task-related streams of information.

With regard to central processing operations, there is some evidence that the performance of two tracking tasks is better if the dynamics on both axes are the same than if they are different, even if the like dynamics are produced by combining two more difficult tasks (Chernikoff, Duey, and Taylor 1960). Even when the performance of two identical but difficult tasks is not actually better than the performance of a difficult–easy pair, performance of the difficult but identical pair is less degraded than would be predicted by a pure resource model (Braune and Wickens 1986; Fracker and Wickens 1989). That is, there is an advantage for the similarity of two difficult dynamics, which compensates for the cost of their increased difficulty.

A similar phenomenon has been observed in the domain of speeded decision making by Duncan (1979). He observed better time-sharing performance between two incompatibly mapped response-time tasks (e.g., left stimulus to right response) than between a compatible and an incompatible one in spite of the fact that the average difficulty of the incompatible pair was greater. Here again, the common rules of mapping between the two tasks helped performance. A related series of investigations has demonstrated superior time-sharing performance of two rhythmic activities when the rhythms are the same rather than different (Klapp 1979; Peters 1981). Investigators have also noted that when manual and vocal response are redundantly mapped to a single stimulus (i.e., both responses are based on the same information), then the bottleneck normally associated with simultaneous response selection is eliminated (Fagot and Pashler 1992; Schvaneveldt 1969).

These examples illustrate that similarity in information-processing routines leads to cooperation and facilitation of dual-task performance, whereas differences lead to interference, confusion, and conflict, an issue we now address.

Confusion

We have discussed ways in which increasing similarity of processing routines can bring about improved dual-task performance. A contradictory trend in which the increasing similarity of processing material may reduce rather than increase time-sharing efficiency is a result of confusion. For example, Hirst and Kalmar (1987) found that time sharing between a spelling and mental arithmetic task (i.e., letters and digits) is easier than time sharing between two spelling (i.e., both letters) or two mental arithmetic (i.e., both digits) tasks. Hirst (1986) showed how distinctive acoustic features of two dichotic messages, by minimizing confusion, can improve the operator's ability to deal with each separately. Many of these confusion effects may be closely related to interference effects in working memory. Indeed, Venturino (1991) showed similar effects when tasks are performed successively so that the memory trace of one interferes with the processing of the other. It is such confusion that will cause greater disruption when trying to do math while listening to basketball scores (i.e., confusing digits and digits) than while listening to a story (i.e., less confusing digits and words).

Although these findings are similar in one sense to the concepts underlying multiple resource theory (greater similarity producing greater interference), it is probably not appropriate to label the elements in question as *resources* in the same sense as the stages, codes, and modalities of Figure 8.1. This is because such items as a spelling routine or distinctive acoustic features hardly share the gross anatomically based dichotomous characteristics of the dimensions of the multiple resources model (Wickens 1991). Instead, it appears that interference of this sort is more likely based on confusion, or a mechanism that Navon (1984) and Navon and Miller (1987) labeled outcome conflict. Responses, or processes, that are relevant for one task are activated by stimuli or cognitive activity for a different task, thus producing confusion or cross-talk between the two. The most notorious example of this phenomenon is in the Stroop task, discussed in chapter 6, in which the semantic characteristics of a color word name interfere with the subjects' ability to report the color of ink in which the word is printed. The necessary condition for confusion and crosstalk to occur is high similarity. That is, *color*—its semantic or visual expression—enters into both the interfering and disrupted tasks. Stroop effects are not found, or are greatly reduced, when people try to report the color of noncolor words (Klein 1964). Though the Stroop task represents a failure of focused attention rather than divided attention, this explanation can also well account for similarity-induced confusion in the latter case.

In summary, although confusion due to similarity certainly contributes to task interference in some circumstances, it is not always present nor always an important source of task interference (Fracker and Wickens 1989; Pashler 1998). Its greatest impact probably occurs when an operator must deal with two verbal tasks requiring concurrently working memory for one and active processing—comprehension, rehearsal, or speech—for the other or with two manual tasks with spatially incompatible motions. In the former case, similarity-based confusions in working memory probably play an important role.

Conclusion

In conclusion, this chapter and the preceding one have generated a growing list of things that can affect the efficiency with which two or more tasks can be time shared—that is, the efficiency of divided attention between tasks. Chapter 7 discussed the obvious candidate of single-task difficulty, draining resources that would otherwise be available for the concurrent task. The present chapter described the similarity of demand for global structural resource and the similarity between tasks of both mappings (i.e., more similarity helps via cooperation) and material (i.e., greater similarity hurts because of confusion). We can think of the mechanisms discussed in this chapter, then, as emergent features that grow out of the relation between the two or more time-shared tasks but that are not properties of either task by itself.

However, as noted already, predicting such dual-task interference, complex as it may be with the five mechanisms of difficulty, multiple resource similarity, preemption, confusion, and cooperation is still only part of the picture. The issue of which task suffers more when there is interference and which is preserved, is determined by the overall task management strategy of resource allocation. One example of such a resource-allocation effect we described briefly was auditory preemption, favoring a discrete task that is delivered auditorally over one delivered visually. Another referred to in the opening anecdote is the compelling nature of certain tasks. Indeed, task management is such a critical aspect of attention that it is addressed at different levels in the following chapter, in which overall task or workload management strategies are described, as well as the concept of executive control, which implements these strategies.

9

Executive Control:
Attention Switching, Interruptions, and Task Management

Introduction

Aviation accidents are often the result of poor task management (Dismukes and Nowinski 2007); the operator switches attention from critical tasks of airplane guidance and stability control to deal with an interruption (e.g., a communication from air-traffic control; a possible failure of landing gear) and then fails to bring attention back to the high-priority safety-critical task. In 1991 in Los Angeles, an air-traffic controller positioned a plane on an active runway, switched attention to a number of unrelated items, and then failed to return attention to the vulnerable airplane and move it to a different runway. Another plane was then cleared to land on the runway where the first plane had been left. Several fatalities resulted from the ensuing crash. In 1987 in Detroit, pilots configuring the airplane for takeoff switched attention to address a request from Air Traffic Control and then returned attention to the checklist-guided preparation activities after missing the critical step of setting the flaps, which were necessary to gain adequate lift on takeoff (Degani and Wiener 1993). In the resulting crash, more than 100 lives were lost.

These are examples of breakdowns in selective attention—the attention element described in chapter 4. However, the present chapter refers to attention directed to tasks rather than to perceptual channels, and the topic thereby can be relabeled as *task management* (Adams, Tenny, and Pew 1991; Damos 1997; Dismukes and Nowinski 2007; Dornheim 2000; Funk 1991; Wickens 2003). Also, in contrast to chapters 7 and 8, where the concerns with task management were those of the allocation of resources during parallel processing activities, here the focus is very much on sequential activities, in which parallel processing either does not or cannot take place.

In many activities, the environment may seem to dictate behaviors and tasks to the operator. This can occur because the appropriate response to a signal of some form is largely reflexive or because it has been inculcated by experience—as, for instance, with the tendency to orient toward the source of a loud noise or the tendency to brake in response to a red light, respectively. In other cases, however, the environment may present a range of potential behaviors, none more urgent or important than another in any perceptually obvious way. In such cases, the operator may easily fail to select or prioritize tasks optimally. After sitting down at the computer, for example, we may choose any of a large number of tasks to perform. Ideally, we might open a word-processing document to carry on writing an unfinished paper. Too often, we may instead double-click a Web browser and proceed to surf the Internet or might set to work on the paper and find ourselves distracted by a seemingly constant arrival of new e-mail and phone calls. The potential result in either case may be that a deadline for finishing the paper is missed. As shown already, the importance of appropriate task prioritization is higher still in dynamic and complex environments such as aviation or the hospital operating room, where the status of the system changes rapidly, the number of tasks to be juggled is large, and the consequences of poor management can be fatal (Chou, Madhavan, and Funk 1996). The typical nurse may have as many as ten tasks in a queue waiting to be performed, and the delay of some of these could have serious consequences for patient safety (Wolf et al. 2006).

This chapter first describes some of basic research on executive control and task-switching, processes that underlie the metatask of task management; then it turns to more applied research that deals with this issue in complex real-world domains and places particular interest in recent work on the psychology of interruptions.

Executive Control

When faced with variety of potential behaviors, how are we able—at least sometimes—to willfully choose to perform those that are most urgent or important and to suppress those that are not? How do we manage, likewise, to suspend or abandon an ongoing behavior when an alternative task assumes a new, higher priority? Models of cognition typically assign the intentional management of thought processes and behaviors to an executive or supervisory attentional component (Baddeley 1986; Norman and Shallice 1986). Here, the terms *executive attention* and *supervisory attention* are used interchangeably. The Norman-Shallice model illustrates the role of supervisory attention well. In this account, a repertoire of cognitive and motor behaviors—routine thought processes and actions that the operator is capable of performing—exists as a set of programs or schemas in long-term memory. A given activity is performed when its schema is triggered. Because many behaviors are mutually incompatible, only one or at best a small number of congruous schemas should be allowed to operate at a given time. Schemas are triggered through a competitive process known as contention scheduling. Here, individual schemas receive activation from the operator's perceptual system. Schemas that represent congruous behaviors then facilitate one another whereas schemas that are incompatible inhibit one another. A given schema finally assumes control of behavior when it achieves a dominant level of activation relative to the competing schema.

By itself, however, this process of stimulus-driven contention scheduling explains only the moment-by-moment, bottom-up control of routine behaviors. To allow an influence of top-down, willful processes and the capability for action planning, the Norman-Shallice model incorporates the supervisory attentional system. One role of the supervisory system is to bias the interactions between schema in a goal-driven manner. This entails holding the current goals in working memory and then providing activation to schema that match the operator's goals and suppressing those that do not. Failure to suppress unwanted schema can lead to capture errors or slips (Reason 1990), in which a familiar stimulus triggers a habitual response that is inappropriate under the circumstances. Driving to the store on a Saturday morning, for example, we may unthinkingly take a turn toward the office, carrying out a routine behavior that is triggered by the context. Not surprisingly, capture errors become more common when the executive attentional system is burdened (Roberts, Hager, and Heron 1994). Additional functions of the executive attentional system are to generate novel patterns of behavior (Baddeley 1986, 1996) or to plan extended behavioral sequences (Shallice and Burgess 1993). Under high levels of cognitive load, therefore, when the executive system is heavily burdened, behaviors tend to become less flexible and more stereotyped (Baddeley 1986). Damage to the executive system, moreover, impairs the ability to preplan sequences of behaviors needed to carry out many complex tasks (Shallice and Burgess 1993).

The operation of executive attentional processes is well illustrated within the real-world situation where an operator engaged in an ongoing task is interrupted by a second task, as happened in the Detroit crash. Either the need to switch attention to the second task may be announced by a signal to the operator, or the operator may decide perform the task wholly of his or her own volition, without an explicit cue to do so. In either case, the operator will be required to suspend the first task and switch attention to the second, a process that in and of itself can consume several tenths of a second (Monsell 2003). While performing the new task, however, the operator must maintain the goals of original task so that it can eventually be resumed (Altmann and Trafton 2002). Finally, when the interruption has been dealt with, the operator should be able to switch attention back to the original task, picking up as fluidly as possible where it was left off.

Task Switching

To study task management at the smallest time scale, we can examine the cognitive mechanics of switching attention from one task to another. Remarkably, even this simple process can entail a substantial time cost, a fact first demonstrated by Jersild (1927). Subjects in Jersild's study were presented lists of items and were asked to work their way through each list performing either or both of two different tasks. In some cases, for example, the stimuli were lists of numbers, and the subjects' task was either to add or subtract

18	60	32 + 3
26	COLD	80 − 3
41	44	38 + 3
45	NEAR	50 − 3
73	77	68 + 3
69	TOP	79 − 3

Figure 9.1   Stimuli like those used by Jersild (1927) and Spector and Biederman (1976). Subjects proceed down each list as fast as possible, performing a mental operation on each item, or trial, in sequence.

three from each item on the list (Figure 9.1, left column). In pure blocks, the subject performed the same task on each item in the list. In mixed blocks, the subject alternated back and forth between two different tasks while working through the list. Jersild found that the time necessary to complete the mixed blocks was substantially longer than the average time needed for the corresponding pure blocks. That is, the need to alternate back and forth between different tasks imposed processing demands beyond those associated with the mathematical operations themselves. Jersild's experimental procedure has become known as the task-switching paradigm, and the response-time (RT) increase produced by the alternation between tasks has become known as a switch cost. Switch costs can be measured by examining trial-by-trial RTs for blocks in which tasks alternate in pairs (e.g., A, A, B, B); the switch cost is the difference between RT following a switch and RT following a task repetition (Figure 9.2). In addition to the trial-by-trial switch costs, RTs in mixed tasks blocks may also show a more general mixing cost.[*] This can be measured by comparing RTs for task repetitions in mixed blocks with the mean RTs for pure blocks. Frequently, RTs for repetitions in the mixed blocks are longer than pure block RTs, indicating an additional slowing of mixed block performance even after trial-by-trial switch costs are accounted for (Kray and Lindenberger 2000; Monsell 2003).

What is the cause of these switch costs? Data indicate that multiple effects contribute. One is uncertainty about which task to perform on a given stimulus. Jersild (1927), after discovering the task-mixing effect, demonstrated that the costs of task mixing were eliminated when the stimuli for the alternating tasks were mutually incompatible so that the nature of the stimulus implicitly defines the nature of the task—for example, when one task was to

---

[*] In the basic attention literature, switch costs as described here are sometimes referred to as local or specific switch costs, whereas mixing costs are referred to as global or general switch costs.

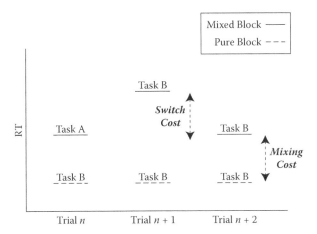

Figure 9.2 Hypothetical data illustrating switch costs and mixing costs. A switch cost is the difference between RT for a given task following a task alternation and RT for the same task following a repetition. Here, the switch cost is the difference in RT for task B on trial *n* + 1 and trial *n* + 2 of the mixed trial block. A mixing cost is the difference between nonswitch RTs for a given task in the mixed block and RTs for that task in the pure block.

add three to a two-digit number and the second task was report the antonym of a common word (Figure 9.1, middle column). An experiment by Spector and Biederman (1976) replicated this effect and also found that the effects of mixing addition and subtraction within blocks were reduced when a +3 or –3 were placed alongside each stimulus as a cue to indicate which task should be performed (Figure 9.1, right column). Spector and Biederman concluded that uncertainty about which task to perform on a given stimulus was one source of the switch cost, suggesting that the cost will be attenuated to the extent that an external cue is available to indicate which task should be performed on a particular stimulus. Thus, switch costs are greatest when stimuli are compatible with either task and when no cue is provided to signal which task is appropriate (Figure 9.1, left column), are reduced when a disambiguating cue is provide (Figure 9.1, right column), and are minimized when the stimulus unambiguously specifies which task to perform (Figure 9.1, middle column). In the absence of external cues, operators tend to rely on rehearsal in verbal working memory to remind themselves which task to perform on each new stimulus. Thus, when operators are prevented from talking to themselves, either out loud or subvocally, the costs of uncued task switches increase (Baddely, Chincotta, and Adlam 2001). The need to maintain multiple-task sets in memory may also contribute to general mixing switch costs shown in Figure 9.2, since additional memory load would be present even on task repetition trials. In dual-task, divided-attention paradigms discussed in the previous two chapters, this cost is sometimes referred to as a *cost of concurrence*, referred to in chapter 7, Figure 7.5.

The need to remember or determine which task to perform on a give stimulus, however, does not entirely account for task switch costs. A transition from one task to another also appears to necessitate a task set reconfiguration, "a sort of mental 'gear-shifting'" (Monsell 2003, p. 136) that may including changing goals, activating new stimulus-response mappings (Rubinstein, Meyer, and Evans 2001), and adjusting the parameters of subordinate perceptual and attentional processes (Gopher, Armony, and Greenshpan 2000; Logan and Gordon 2001). This reconfiguration accounts for at least part of the specific switch cost. The time needed for reconfiguration can be estimated using a procedure in which the order of tasks varies randomly and a cue is presented before each target stimulus to indicate which task should be performed. Data from such experiments indicate that switch costs increase with task complexity. It takes longer, for example, to establish the proper mental configuration for a task with difficult stimulus-response mappings than for a task with simpler or more natural mappings (Rubinstein et al. 2001). Conversely, switch costs tend to decrease as the interval between the cue and target grows longer. However, even when the operator is given a long preparation period, the switch cost is not entirely eliminated. These results suggest that the operator can begin task-set reconfiguration when cued but that the process cannot be completed until the target stimulus arrives to provide an exogenous event-driven trigger (Meiran 1996; see also Rogers and Monsell 1995).

Because task-set reconfiguration is a responsibility of the executive attentional system, it is hindered by a secondary task that also burdens executive attention processing—for example, a task that requires the juggling of information in working memory or retrieval of information from long-term memory (Baddeley et al. 2001). Switch costs can likewise be inflated by interference from carryover activation of a task set that is no longer appropriate, or by carryover inhibition of a set that was previously inactive but is now required—effects that have been labeled task-set inertia (Allport, Styles, and Hsieh 1994). The phenomenon of task-set inertia implies that even after an interrupting task has been completed, the act of having performed it may continue to hinder the ongoing task, contributing to general switch costs of the form just described. Such an effect parallels the well-known effect of proactive interference in memory, in which earlier but no longer relevant material continues to interfere with the ability to learn and to remember later arriving material.

Importantly, the phenomenon of switching costs scales up very nicely from the basic laboratory research to more applied environments. For example, Wickens, Dixon, and Ambinder (2006) observed relatively large (> 1 second) costs as pilots switched between subtasks of controlling and supervising two simulated unmanned air vehicles. These costs were inferred from the difference in task times when completed singly and in combination, much like the psychological refractory period research described in chapter 2.

Task Interleaving

Switch costs are but one, albeit important, component of task management, when divided attention between two or more tasks, involves switching back and forth between them—that is, interleaving between two tasks. A related phenomenon concerns the circumstances of return back to one task (A) after having dealt with another (B). Returning to the interrupted task, the operator may resume performance at the point where it was interrupted or may pick up the task at an earlier or later point. For example, a skilled musician who is interrupted in midphrase might return to the beginning of a phrase or perhaps even to the beginning of the piece when resuming his or her performance. Alternatively, as in the tragic aircraft accident in Detroit described at the outset of the chapter, an operator might resume performance of the interrupted task beyond the point at which it was suspended, omitting an intended step. Importantly, the double switch (A→B→A) may be a natural part of time sharing, or time swapping, two intended tasks, as switching gaze in a repeated cycle between the roadway and a head-down display; alternatively, B may be a specific event-driven interruption (Trafton 2007; McFarlane and Latorella 2002). Finally, earlier we spoke of returning eventually to the original task. However, sometimes this return may never take place at all, which is considered a failure of prospective memory (Dismukes and Nowinski 2007; Harris and Wilkins 1982; McDaniel and Einstein 2007), a form of memory that describes remembering to do something in the future.

A context for discussing the research findings on task interleaving is presented in the ongoing-interrupting task (OT-IT) diagram shown in Figure 9.3. Here the operator is performing some ongoing task, shown by the two OT boxes at the top of the figure in panel A. This task is interrupted by an interrupting task. The arrival of the IT may or may not be announced, and in either case it may take some time before the operator disengages from the OT to initiate the IT—or, of course, the operator may ignore the IT entirely, a strategy not shown in Figure 9.3. Once the operator leaves the OT, there may be some attention-switch time before the IT can be performed; this is called switch 1. Correspondingly, after completion of the IT, there may be an attention-switch return time before the OT can be resumed; this is called switch 2. Trafton (2007) referred to these two times as the interruption lag and the resumption lag, respectively. The amount of time that the IT is performed is analogous to the dwell duration discussed in chapter 4—the delay in return being analogous to the first passage time. Once attention is returned to the OT, there may be an initially degraded quality of performance, as it may take time to get engaged again if we forgot where we left off the OT when we departed, or we may make mistakes on return that are a carryover from the IT. Finally, at the bottom of the figure in panel B, we show the total time to do the OT within the dual-task, or task-switching, context and compare this with the time it would have taken to complete the OT in single-task conditions. The difference is the cost of interruptions, a cost that has been

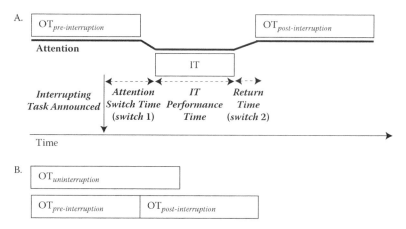

Figure 9.3 Effects of an IT on performance of an ongoing task OT. Panel A presents the time course of task performance, with the switch from and back to the OT. The total time needed to perform the interrupted OT includes the time spent on the task preinterruption; the time necessary to switch attention from the OT to the IT, to complete it, and to switch attention back to the OT; and then the time needed to resume and complete the OT. As depicted in panel B, the total time needed to perform the OT itself may increase under interrupted relative to uninterrupted conditions even when excluding attention switch times and IT task performance. This increase is a result of the time needed to resume the OT after returning from the IT.

well documented in the literature (see Trafton 2007). So, too, has been documented the frequency of interruptions and task switching in the workplace such as the aircraft cockpit (Dornheim 2000). Gonzales and Mark (2004), for example, found that task switching of information workers took place about every three minutes, and Wolf et al. (2006) observed that nurses were interrupted in their activities at least every twenty minutes. McFarlane and Latorella (2002) describe a large number of factors that influence the fluency of interruption management.

As noted already, the representation in Figure 9.3 can be applied in one of two contexts. On one hand, it may characterize the interleaving of two relatively ongoing tasks so that the distinction between IT and OT is somewhat arbitrary (e.g., the cycle shown in Figure 9.3 continues); this may characterize map checking while driving. On the other hand, it can characterize the response to a particular single interruption, such as a phone call while working on a word processor. The following section describes the effects on this process from the standpoint of the second of these contexts, recognizing that this analysis can generalize to the first as well.

Interruption Management

One can analyze the processes in panel A of Figure 9.3 in terms of four sequential task properties: those that influence attention switch 1 (interruption), that are characteristic of the OT and the IT, and those that effect switch 2 (return) that are properties of the IT and the OT. Each of these is described in turn.

Switch 1: OT Properties

Three factors directly influence the likelihood and speed of leaving the OT to deal with an IT—or, in the interleaving case, the tendency to stay longer on an OT before switching.

Engagement and Cognitive Tunneling on the OT

Though it is intuitively obvious that a task will be less susceptible to interruption when the operator is highly engaged in it, this effect has been somewhat difficult to capture experimentally or parametrically for the purposes of modeling. Interest plays a role. Interesting tasks are engaging, and we are reluctant to leave them, just as boring tasks can be easily interrupted. The role of interest is revealed by the results of a recent meta-analysis of research on cell-phone disruption of driving performance. Horrey and Wickens (2006) found that studies using actual conversations tended to disrupt driving more than those simulating the information-processing demands of such conversations with cognitive tasks but that were less engaging. That is, the latter tasks did not involve interesting semantic content, whereas the former were generally explicitly designed to attract the interest of the participant (e.g., discussing current topics, personal stories).

In addition to interest, at least two other sources of cognitive tunneling can be identified. First, the phenomenon has been observed in highly immersive realistic three-dimensional displays (Wickens 2005b). For example, a form of cockpit navigational display known as the three-dimensional highway in the sky (Alexander, Wickens, and Hardy 2005) was found to engage headdown attention to a degree that pilots sometimes failed to notice critical events that took place in the outside view of the world but that were not rendered on the engaging three-dimensional virtual reality display in the cockpit. Such failures were not observed when pilots flew with more conventional two-dimensional flight instruments (Wickens 2005b). Second, cognitive tunneling has also been observed during critical problem-solving or trouble-shooting operations, with people failing to notice other important events (Dismukes and Nowinksi 2007; Moray and Rotenberg 1989). This phenomenon was manifest in the Eastern Airlines Everglades crash in 1972, when pilots, trying to diagnose a landing-gear problem, were entirely unaware of a salient auditory alert that signaled their impending crash into the ground.

Strategic Factors and the Stopping Point

Operators may choose to remain with an OT for a while before switching until they get to a stopping point, perhaps as a way of resolving some ongoing subtasks and thereby of avoiding the need to hold information in working memory during the IT period or avoiding the need to reacquire forgotten information when the OT is resumed. Altmann and Trafton (2002) and McFarlane (2002) modeled the OT in terms of hierarchical goal structures, noting that the OT is more likely to be left after a goal is completed than in the middle of a completion sequence (e.g., carrying out a sequence of programming subtasks that lead to a task goal, like finishing writing a sentence or paragraph). The OT is more rapidly resumed on return when it was left at a good stopping point (e.g., between tasks or subtasks) than in the middle of a subtask (Bailey and Konstan 2006; Monk, Boehm-Davis, and Trafton 2004). It is also better resumed when a delay is taken after the interruption and before the OT is left—for example, a large switch 1 or interruption lag (Dismukes and Nowinski 2007).

Presumably part of the benefit of this delay—the interruption lag—is that it allows people to better encode the state of OT into memory before leaving it so that the point of departure will be well remembered and, therefore, the point of return easily found (Trafton, Altman, and Brock 2005). In this light, it is noteworthy that when people intentionally take a longer time with switch 1, this strategy of prolonging switch 1 will have beneficial effects on return, giving people time to think about where they left off or possibly placing a visible reminder in the OT workplace (Dismukes and Nowinski 2007). Also, designers of computing systems are considering ways in which human cognitive operations can be monitored to adaptively impose interruptions only at optimal stopping points during ongoing tasks (Bailey and Konstan 2006).

A closely related task characteristic found to influence switching strategy is the working memory demand of the OT. For an auditory working memory task, like dialing a long phone number just heard on an answering machine, there should be reluctance to leave it until the dialing task is completed because some of the dialed digits may have been forgotten after an interruption-induced switch. This would not be the case when dialing from a visual phone number on a device with visual feedback of the dialed digits. Here it is noteworthy that visual—rather than auditory—presentation of complex information allows better task management and more optimal task switching (Wickens and Colcombe 2007b) because with a permanent visual text display of an OT, switch 1 can occur without fear that the information will be gone on return. This is, of course, not the case with working-memory-challenged auditory information.

In dynamic-process or vehicle-control tasks, operators should be more reluctant to abandon the OT control task if the systems involved are high bandwidth or unstable; they also may choose to switch at moments of time when the system is most stable (e.g., the car is in the center of the lane and on a forward trajectory, and the road ahead is straight). In this case, how much neglect will lead to a significant state change depends on the inertia and dynamics of the system (Wickens 1986, 2003). A large aircraft flying in still air will allow longer neglect than a light aircraft flying in turbulence. Furthermore, some tasks, like flying a helicopter or riding a bicycle, are inherently unstable and lead to time-dependent divergence if neglected. Finally, it should be noted that the consequences of not following a deferred switch strategy and therefore of abandoning the OT (switch 1) in the midst of a goal pursuit (subtask), of a working memory-loading operation, or in an unstable dynamic situation will be realized on return to the OT, as discussed below (Monk, Boehm-Davis, and Trafton 2004).

Importance and Priority

Like engagement, it is also intuitive that an OT with higher priority than an IT should sustain its performance longer in the face of an interruption and generally be less interruptable than a task of lower priority. This effect was found by Iani and Wickens (2007) as a primary flight task was interrupted by the delivery of discrete weather information. Turning to ground transportation, the fact that driving is as safe as it is—despite all of the multitasking that goes on, much of it head down—suggests as well that people's switching strategies tend to prioritize out the window viewing over head-down activity (Wickens and Horrey in press), as if the out-the-window tasks are less interruptible. However, evidence that such prioritization does not always hold comes from the numerous examples in which this sampling strategy fails and in which accidents occur as a result of in-vehicle distractions (Dingus et al. 2006). One important issue addressed in chapter 3 is the ability of operators to know the priority of the IT before the full interruption has taken place and the OT has been abandoned. This points to the importance of preattentive referencing in alarms (Woods 1995) as discussed in chapter 3, and systems that contain preattentive referencing have been found to be effective in task management (Ho et al. 2004).

Switch 1: IT Properties

Importance

As with the OT properties, the IT importance is again a relevant factor for switch 1, as it is indeed the relative importance between the two tasks that matters most. The less important IT will prolong switch 1. The role of relative importance in governing scanning between head-up and head-down driving tasks was well documented by Horrey, Wickens, and Consalus (2006), and chapter 4 showed for visual tasks the prominent role of task value, or importance, in the SEEV model.

Salience

As chapter 3 demonstrated, the salience of stimuli and events that accompany the IT will prominently influence the switching speed. In particular, auditory events are typically more salient than visual ones and will lead to faster switching speeds, a phenomenon referred to as the *auditory preemption* effect (Ho et al. 2004; Wickens and Colcombe 2007b; also see chapter 8 in this volume)—although this effect is not always observed, particularly when visual events have direct access to foveal vision. Another distinction of IT salience is between what we call announced and unannounced interrupting tasks. The announced tasks have a stimulus event associated with it—typically visual or auditory—and these are clearly more salient than unannounced ones in which IT initiation must depend on some form of prospective memory. The unannounced tasks, like remembering to turn the heat down before the pot boils over rather than afterward, are more likely to get missed or delayed.

Switch 2: IT Properties

The case can be easily made that properties of the IT that will lead people to delay switching back to the OT are very much the same properties that lead people to stay on the OT prior to switch 1, as discussed previously. Indeed, there is a modest, but imperfect, reciprocity between IT and OT properties in their influence on switching performance (Iani and Wickens 2007).

Switch 2: OT Properties.

Finally, we address characteristics of the OT that influence the circumstances of return to it after switch 2.

Strategies that Were Carried Out at Switch 1

Here research indicates that the most important variable is that associated with the strategies adopted on leaving the OT for switch 1, as discussed already, and there are several of these. For example, to the extent that OT was left in the middle of a subtask—rather than between subtasks—the OT is more likely to be degraded on return (Miller 2002; Monk, Boehm-Davis, and Trafton 2004). Indeed, Miller (2002) found that the interruptions were often so disrupting that some OTs needed to be restarted from the beginning such that the time taken to complete OT postinterruption (Figure 9.3) was fully as long as OT uninterrupted; that is, nothing was accomplished by OT preinterruption—operators needed to start from scratch. As noted already, an important strategy at switch 1 that affects performance in OT resumption is the placement of an intentional delay at switch 1. This delay can be used to accomplish either of the following beneficial acts: (1) It can be used to rehearse the state of the OT at the interruption (Trafton et al. 2003); and (2) it can be used to provide an explicit, usually visual, placeholder such as the mark on the page where text editing was left off when answering a phone call. Both of these actions are helpful (Dismukes and Nowinzki 2007; McDaniel and Einstein 2007).

Delay in Return

Delaying the resumption of the OT can degrade the quality of return in two ways. First, there will be a simple decay of working memory, of where the task was left off; or if the OT had loaded working memory at switch 1, then there will be a decay of the material in the task itself (remember the phone dialing example). Second, if the OT is a dynamic one, like vehicle control, the delay or neglect may increase the likelihood that the system itself will have evolved toward an unstable or undesirable state while attention, and therefore control, was absent. Thus, a car will be more likely to have diverted toward the ditch the longer the head stays down. As described in the context of scanning in chapter 4, a long first-passage time away from the OT leads to increasing vulnerability (Horrey and Wickens 2007; Sheridan 1970). Third, the longer one stays on an IT, the greater is the possibility that the OT will have been forgotten entirely: Its goal memory will have decayed below threshold (Altmann and Trafton 2002), and a failure of prospective memory will have occurred (Dismukes and Nowinski 2007; McDaniel and Einstein 2007).

IT–OT Similarity

Finally, as a property not associated exclusively with either the IT or the OT, Gillie and Broadbent (1989), Dismukes and Nowinksi (2007), and Cellier and Eyrolle (1992) all documented the degrading role of similarity between OT and IT on the resumption of the OT. This effect appears to reflect the same information-processing mechanisms described in chapter 8 related to confusion and cross-talk. Greater similarity between the OT and the still-active IT will cause greater confusion and interference from the still-active traces of the IT when the OT is reinitiated (i.e., proactive interference) and greater disruption of relevant OT information that needed to be retained during the IT period (i.e., retroactive interference). This will be particularly true if the OT required retention of material (i.e., rehearsal) during the OT1–OT2 interval given the resource demands of working memory.

Task and Workload Management

The area of strategic task or workload management integrates the switching and interruption findings discussed already and places them in a broader context. This context might be represented as zooming out away from the single OT→IT→OT element to consider lots of tasks in sequence so that the distinction between OT and IT is blurred. Historically, this approach received a major boost when Hart (1989) and Hart and Wickens (1990) noted that all of the extensive work being done on measuring mental workload (see chapter 7) and evaluating multiple-task interference through resource models failed to account for how people managed multiple tasks when workload was so excessive that concurrent processing was impossible. Thus, research focus was required on how people deal with these overload situations.

Freed (2000) proposed a reactive prioritization model to account for task management with similarities to Sheridan's (1970) earlier modeling on supervisory sampling of input channels (for a good review, see Moray 1986; see also chapter 4 in this volume). Freed (2000) considered four factors that should optimally influence decisions to switch between tasks:

(1) Urgency: How long is it until the deadline by which a task must be completed, and how long does it take to complete the task? For example, five minutes remaining until the deadline of a four-minute task has an urgency of one minute. If one minute passes and the task is not initiated, it will be too late to finish it on time.

(2) Importance: What is the cost of not doing the task? It is acceptable not to switch to an urgent (factor 1) task if there is no cost in missing its deadline. This factor will lead to a greater proportion of time spent doing more important tasks (Raby and Wickens 1994).

(3) Duration: Longer-duration tasks will, of course, increase the urgency if not yet performed, but they will also be more likely to disrupt performance of other tasks once they are initiated, assuming that there is a task-switching cost to leaving them temporarily uncompleted (factor 4). This penalty will make operators reluctant to leave the longer-duration task.

(4) Switching or Interruption Cost: This concept has been discussed repeatedly already. A high switching cost will lead to task inertia and a likelihood of continuing without a switch (Tulga and Sheridan 1980; Ballard, Hayhoe, and Pelz 1995). Note, by the way, how this cost of switching is tied to the concept of information-access effort or cost, as discussed in chapters 4 and 7 in this volume. Greater distance between visual sources of task information will lead to greater costs of switching.

Added to this mix is, of course, the role of uncertainty. Sometimes we do not know before switching how long it will take to do the task once initiated, nor do we always know about the impending arrival of additional tasks. In the latter regard, Tulga and Sheridan (1980) demonstrated the value in optimal task management of knowing in advance (i.e., preview) what arriving tasks will be and how long they are likely to take.

Freed's (2000) model does not appear to have been fully validated regarding the extent to which people follow its optimal prescriptions or the circumstances that make one factor dominate over others. Tulga and Sheridan (1980) did provide some data, but only using very generic pseudo-tasks: computer-displayed bars that can be activated to simulate attention directed to the bars. Raby and Wickens (1994) performed an empirical task-management study with trained airplane pilots to examine some aspects of optimal scheduling. Their pilots flew a simulated approach in a realistic airplane simulator under three conditions of increasing workload, varied by the amount of time pressure on the pilots to complete all the tasks necessary to plan for and to accomplish a landing at an unfamiliar airport. Prior to the study, the nineteen tasks to be performed were categorized into three categories of priority, or importance, labeled *must, should,* and *can.* These corresponded somewhat to the more generic aviate-navigate-communicate-system management task importance hierarchy traditionally used in aviation (Schutte and Trujillo 1996). The investigators then evaluated the overall flight quality (i.e., precision of flying) as well as the specific timing and performance of the three categories of tasks. The results revealed that pilots generally behaved appropriately, as characterized by the following:

- Performing tasks at more optimal times, prioritizing must tasks over should tasks, and prioritizing should tasks over can tasks
- Abandoning, or shedding, can tasks more frequently than should tasks as workload increased and abandoning should tasks more frequently than must-do tasks

However, Raby and Wickens (1994) found that when workload increased, pilots did not not optimally reschedule the higher-priority tasks in response to the dynamic workload change. From this they concluded that pilots do not maintain perfectly optimal strategies for the plausible reason that the task scheduling itself demands resources that should otherwise be devoted to performing the tasks themselves. Such a conclusion is consistent with Kahneman's (1973) effort-conserving view of heuristics (chapter 7 in this volume) and with the results of another scheduling study carried out by Moray et al. (1991).

Another aspect of Raby and Wickens's (1984) study examined differences in task-switching behavior between the better-performing and worse-performing pilots. Three conclusions emerged here regarding better-performing pilots, both also supported by other research:

(1) They tended to be more proactive, initiating high-priority tasks earlier (see also Laudeman and Palmer 1995; Orasanu and Fischer 1997). Procrastination hurts. However, too much early preparation in an uncertain world can be counterproductive, especially if formulated plans are rigidly maintained despite changing circumstances. This issue of the plan continuation error is beyond the scope of the current chapter (McCoy and Mickunas 2000).
(2) They tended to switch attention more rapidly between tasks—hence being less inclined toward cognitive tunneling or task inertia.
(3) They tended to be more flexible in attending to tasks as their priority may momentarily change (Schutte and Trujillo 1996).

Conclusion

In conclusion, executive control and task management play the dominant role in multiple-task performance, when parallel processing is no longer possible and when serial processing predominates. This parallels the role of the resource-allocation policy, discussed in parallel processing in the previous two chapters. There is some convergence in results describing these task-management processes when we look at the three domains of research covered here: (1) basic attention switching; (2) more complex interruption management; and (3) the complex level of real-world task and interruption management. However, more research in the third domain is certainly needed to better understand task-management strategies here. All three domains point to the costs of attention switching and interleaving compared with staying on a single, uninterrupted task. All point to the role of strategies and working memory in managing these multiple tasks well.

Also emerging from the literature are findings that certain task- and display-related factors *drive* task management in nonoptimal ways, analogous to the nuisance properties of salience and effort in the SEEV model discussed in chapter 4, whereas other, more optimal, top-down properties can be achieved by the well-calibrated task manager. More optimal task management is possible to the extent that the operator has the information and knowledge available to know things like calibrated task importance, expected value (risk of failing to do the task), duration, and arrival time. This knowledge is analogous to the E and V properties of the SEEV model. In particular, this latter class of top-down knowledge-driven factors in task management points to the possible role of attention and task-management training in improvement of multitasking, resource allocation, and interruption management (Dismukes 2001; Gopher 1992; Hess and Detweiller 1994), an issue addressed in the next chapter.

10

Individual Differences in Attention

Introduction

It has become somewhat of a cliché that we have entered the information age, but it is certainly true that that many of us find ourselves in an age of information overload in which we are bombarded by computer messages, cell-phone calls, PDA messages, and so forth—in which we are constantly multitasking. Some have speculated that such constant exposure to multiple electronic channels improves our overall multitasking ability through practice and experience, and others speculate that younger generations, growing up within such an environment, may possess an improved ability, almost as if it is a trait (Lohr 2007; *Time* 2006). Anecdotal evidence seems to support this speculation. Everyone has some acquaintances who seem to be talking, texting, and listening all at once and all the time and others who have trouble dealing with more than one task at a time.

The current chapter deals with three aspects of these individual differences in attention. First, consideration is given to how differences are created through training and experience. What makes the expert seemingly more proficient at dual tasking than the novice, and can such a skill be explicitly trained to create a shortcut to this important aspect of expertise? Next is addressed the differences between people unrelated to practice and, hence, more attributable to innate individual differences, analogous to verbal and spatial ability differences. What is the evidence for stable traits in the various manifestations of attention we have described, how are these traits assessed, and what differences do these assessments predict? Finally, an examination is given in detail of one very important form of individual difference contributing to attentional performance: biological age, an issue of increasing importance as our aging population strives to remain active and independent on the road and in their homes (Fisk and Rogers 2007; Fisk et al. 2004).

Attention as a Skill

When observing a skilled operator perform a complex task—whether juggling, taking dictation, flying an aircraft, or inspecting products off an assembly line—the novice is often awed at the ease with which the expert time shares multiple activities. Such performance differences between novices and experts could result from any one or more of four bases: (1) single-task skill level; (2) skill automatization; (3) task-specific time-sharing skills; and (4) domain general time-sharing skills. We consider evidence for each of these factors herein.

Single-Task Skill and Automaticity

It is well established that continued practice on a dual task leads to improved performance (Spelke, Hirst, and Neisser 1976). Naturally, some component of this gain could result simply from an improvement in the single-task component skills. Furthermore, the development of automaticity may improve dual-task performance even when no changes in single-task performance are evident. Within the framework of automaticity discussed in chapters 2 and 7 (Figure 7.4), practice that reduces single-task resource demands will free resources to be allocated to a secondary task, thus improving dual-task efficiency (Schneider 1985; Schneider and Detweiler 1988). This may not entail an observable improvement in single-task performance, but only an increase in the data-limited region of a performance resource function (PRF) that asymptotes at the same level (Figure 7.4). As noted already, automaticity develops with extended practice on tasks with consistently mapped components: repeated perceptual elements, co-occurring cues, or stimuli or categories that are consistently mapped to response classes and to physical responses.

Because this form of improvement does not reflect the acquisition of a time-sharing skill but rather a reduction in single-task resource demands, it can be achieved through single-task practice. Such a mechanism explains the automatic processing of familiar perceptual stimuli such as letters (LaBerge 1973), consistently mapped targets (Schneider and Fisk 1982; Schneider and Shiffrin 1977), and repeated sequences of stimuli and events (Bahrick and Shelley 1958; Bahrick, Noble, and Fitts 1954) as well as the automatic performance of habitual motor acts such as signing one's own name, tying a shoe, or logging onto e-mail. When discussing response execution, this automaticity of performance is referred to as the *motor program* (Keele 1973; Summers 1989). Differences in time sharing capabilities between flight instructors and student pilots (Damos 1978) can easily be explained by the greater automaticity with which the former carry out many aspects of their task.

Specific Time-Sharing Skills

In contrast to automaticity, strong evidence for the existence of time-sharing skills falls from the following logical arguments:

(1) Single-task automaticity develops best when full resources are allocated to learning the components (Lintern and Wickens 1991; Nissen and Bullemer 1980).

(2) If single-task automaticity were the only element underlying the high time-sharing, or divided-attention, competence following practice, then single-task practice of components—which in training research is called part-task training, or fractionation (Wightman and Lintern 1985)—would be the most efficient way of improving time-sharing performance.

(3) A review of the literature on part-task training suggests that such training is often no better and sometimes worse than whole-task training on a dual-task combination (Damos and Wickens 1980; Fabiani et al. 1989; Lintern and Wickens 1991) in spite of the fact that whole-task conditions hamper the development of automaticity. Hence, the existence of a learned time-sharing skill is revealed.

The revelation of a time-sharing skill leads to the next question: What does such a skill consist of? Research has suggested at least four possibilities, all tied to the general notion that the expert learns to strategically allocate attention.

First, time sharing is facilitated by the development of visual scanning strategies in multitask situations. For example, expert pilots learn scanning patterns necessary to control all three axes of flight (i.e., triple tasking) (Bellenkes, Wickens, and Kramer 1997). Skilled vehicle drivers learn to scan farther down the highway (Mourant and Rockwell 1972) and to orient toward sources of potential hazards (Fisher and Pollatsek 2007). As noted in chapter 4, many of these aspects of skills can be associated with a selective attention-scanning strategy that adheres to the expected value component of the SEEV model—that is, a calibrated mental model (Moray 1986). Thus, Wickens, McCarley, et al. (2007) found that superior performers in a multitask flight simulation had scanning strategies that approximated those prescribed by the optimum expected value model, unlike the participants who fared more poorly.

Given that scanning differences underlie proficiency differences in dual-task performance, one can ask if such differences can be explicitly trained. Here the research of Pollatsek et al. (2006), which was described in chapter 4, is promising. This work trained novice drivers to better scan to high-risk areas in the forward field, thus allowing better performance on the dual tasks of lane keeping and hazard monitoring.

A second source of time-sharing skill can be understood within the framework of the task-switching and interruption literature described in the previous chapter. There, mechanisms were discussed that improve task and interruption management; in particular, an experiment was noted by Dodhia and Dismukes (2003) and Dismukes and Nowinski (2007), which demonstrated that resumption of an ongoing task following interruption was easiest if (1) there was a delay allowed between the interruption and the suspension of the ongoing task, allowing the performer to choose a convenient leaving place and perhaps to place reminders of where to return to the ongoing task; and (2) when the interrupting task was completed there was some salient reminder existed to return to the ongoing task, perhaps one planted in (1).

It seems reasonable that such strategies—as well as others (McFarlane 2000; McFarlane and Latorella 2002) such as being aware of when interruptions are particularly insidious (McDaniel and Einstein 2007)—could be taught as a way to improve task management. To date, little data exist regarding the value of such explicit training in transfer to new tasks. Nevertheless, at least one study (Hess and Deitweiler 1994) established the value of mere exposure to interruptions as a way of developing interruption-handling skills, whereas Cades, Trafton, and Boehm-Davis (2005) found that explicit training on the interruption–resumption interval could produce improved performance above and beyond simple performance on the tasks themselves. Dismukes (2001) advocated such interruption management training for pilots.

A third source of evidence for time-sharing skill comes from the success of training fairly specific resource-allocation strategies, in which resources are considered in the context of chapters 7 and 8. Strong evidence has supported the success of what is termed *variable priority training* (Gopher 1992, 2007), suggesting that attentional flexibility is indeed an important time-sharing skill. Gopher and Brickner (1980) observed that subjects who were trained in a time-sharing regime that successively emphasized different resource-allocation policies became more efficient time sharers in general than did a group trained only with equal priorities. The former group was also better able to adjust performance in response to changes in dual-task difficulty. Fabiani et al. (1989) and Gopher, Weil, and Siegel (1989) found that training that emphasizes attention control leads to better transfer to a complex multi-task video game, and further evidence suggests that similar training benefits later performance in the attention-challenging task of flying military aircraft (Gopher, Weil, and Bareketi 1994). As discussed following, Kramer, Larish, and Strayer (1995) found that variable priority training could offset some of the deficits in time sharing shown by older adults and that this skill in learning how to flexibly allocate resources could transfer between different dual-task combinations. Schneider and Fisk (1982) found that subjects could time share automatic and resource-demanding letter-detection tasks with perfect efficiency if they were instructed to allocate their attention away from the automatic task. In the absence of this training, subjects allocated resources in a nonoptimal fashion by providing more resources to the automatic task than it needed, compromising performance of the resource-limited task. Thus, clearly, resource allocation can be trained, and such training leads to better multitasking.

A fourth indication of what may underlie time-sharing skills comes from the research of Damos and Wickens (1980), who had subjects practice extensively on dual-task pairs involving both tracking and discrete digit processing. Fine-grained analysis of dual-task performance revealed that practiced subjects engaged in more parallel processing of the stimuli rather than discrete switching. Thus, it is possible that more proficient time sharing involved either developing knowledge that greater efficiency can be gained by parallel processing or that the actual capacity, or size of the resource pool, increases.

The success that researchers have had in training attentional scanning (Fisher and Pollatsek 2007; Pollatsek et al. 2006), visual search (Gramopadhye, Drury, and Prabhu 1997), and task prioritization (Gopher 2007) parallels other efforts to inculcate attentional skills. Walker and Fisk (1995) designed a computer game to train football quarterbacks to reading the defense and were successful enough that their technique was adopted by the Atlanta Falcons. A similar intervention that emphasized attentional flexibility has been employed with success in training college and professional basketball players (Gopher 2007). Navarro et al. (2003) developed a tool to help children divide visual attention between different elements to be compared in complex visual scenes (e.g., "One of these faces is not like the other") and found that benefits from training on this task transferred to improved performance on other cognitive tests that depended on attention. Green and Bavelier (2003) observed that habitual video-game players performed better on tests of attention measuring the useful field of view (UFOV) (i.e., attentional breadth) than did nonplayers. The investigators then assessed whether the UFOV could be expanded for nonplayers by having them practice for eight sessions on the video game, and they found that it could.

Evidence for such things as a broadened UFOV or a practice-induced increase of attentional flexibility suggests that some aspect of dual-task proficiency may be general in nature, applicable not only in the specific task-sharing context in which it was acquired but in other multitask combinations as well. For example, practicing cell-phone talking while driving might transfer to a qualitatively different time-sharing task combination (e.g., rehearsing a lecture while cooking). To date, however, evidence for such general time-sharing skill, learned through training, is slight but still positive. Thus, Damos and Wickens (1980) found that a transfer from dual-axis tracking to dual-digit processing—and vice versa—was positive, suggesting that the earlier dual-task training does, in fact, develop a generalizable multitasking skill. Kramer, Larish, and Strayer (1995) found that, particularly for older individuals, variable priority training on one dual-task combination led to positive transfer on a very different task combination. A related finding is that bilingual children who are raised in a household where attention must often be switched between languages show a pattern of more proficient executive control than children raised in a monlingual household (Bialystok 1999). These pieces of positive evidence notwithstanding, it is probably the case that the greatest component of time-sharing skills lies in learning the effective resource-allocation skills and strategies for a specific dual-task pair. Other more enduring and general differences across task combinations are probably more attributable to stable individual differences—traits of attention—an issue to which we now turn.

Attention as an Ability

As distinct from a skill, which is the result of practice, we consider here the role of ability in attention, in which ability is assumed to be a more innate and permanent cognitive characteristic, akin to the stable differences between people in verbal and spatial abilities. Such individual differences in attentional capability are important because of their potential role in selection of workers to carry out complex attention-demanding jobs, like piloting an aircraft or managing an office. For example, if reliable tests of attention can be shown to correlate with performance on the flight deck, then it should be possible to screen applicants for a highly competitive piloting job on the basis of their test scores (Caretta and Ree 2003; Hunter and Burke 1995; Pohlman and Fletcher 1999). To address the issue of attention abilities, we consider two broad classes of studies: (1) those that have examined stable individual differences in attentional components; and (2) those that have examined overall differences in time-sharing ability (Brookings and Damos 1991). Within each class, researchers' interests have been focused on correlational measures indicating that people who do well on one particular attention test also do well on another or on some complex skill requiring heavy attentional support (Hunt and Lansman 1981).

Attentional Components

Attention switching has received a fair amount of research as an ability component. Much of this research was stimulated by the early findings of Gopher and Kahneman (1971), Gopher (1982), and Kahneman, Ben-Ishae, and Lotan (1973) that measures of auditory attention-switching speed derived from the dichotic listening task discussed in chapter 2 could be used to predict the success rates of student pilots and the accident rates of bus drivers. This test measured how quickly operators could switch their attention on cue from one ear to the other. Research by Lansman, Poltrock, and Hunt (1983) also identified an abilities component related to attention switching. However, the degree of generality of such a component remains uncertain. Lansman and colleagues, for example, found that there were separate components within each the auditory and visual modalities. Correspondingly, Braune and Wickens (1986) failed to find that measures of auditory attention switching correlated with a variety of visual attention-switching measures.

Still, a review of research carried out by Hunt, Pellegrino, and Yee (1989, p. 308) on attention switching both between dual tasks and between components within a single task concluded that there does appear to be "an ability to coordinate information from several sources that is independent of the ability to process information from one of these sources alone." The role of attention control in this process is further supported by the research of North and Gopher (1976), who found that individual differences in the ability to modulate resources to primary and secondary tasks were both uncorrelated with single-task performance and independently predictive of success in flight training. This role of attention control as an ability appears to be closely related to the capacity of working memory, as discussed later in this chapter.

A somewhat related line of research has focused on an emerging distinction between two qualitatively different dual-task processing styles, characterized as parallel processing versus serial processing. The parallel processor, for example, might continue steering while performing a discrete task in the automobile, whereas the serial processor would pause the steering task while dealing with the discrete task or might postpone the discrete task until steering requirements have stopped. This dichotomy, paralleling the contrast between chapters 7 and 8 on the one hand and chapter 9 on the other, seems to distinguish two different classes of people given a dual-task challenge (Braune and Wickens 1986; Damos, Smist, and Bittner 1983). Importantly, it is not always the case that parallel processors show better dual-task performance. As intuition tells us, sometimes it is better to try to do two tasks sequentially and perfectly than to attempt to time share them, suffering decrements on one or both. That is, parallel processing does not necessarily imply perfect performance.

Chapter 5 discussed the UFOV as an important component of visual attention related to visual search. Here Owsley et al. (1998) found that this component can be reliably measured and is predictive of differences in driving safety. They made a compelling case that its assessment should be included in tests of driving competence or selective screening for driver's licenses.

Time-Sharing Ability

An alternative approach to measuring the attention components that often underlie dual-task performance is to directly assess individual differences in dual-task decrements and the extent to which these are stable across people. Of course, here the same care must be exercised as was discussed in the section on attention training. If one person has a smaller dual-task decrement than another, it may simply be because that person possesses greater automaticity on one or more of the component tasks, a difference unrelated to a time-sharing ability. To examine these issues, researchers have tried to disentangle differences in the resources demanded by a component task (i.e., the shape of the PRF) from differences in the resources available (i.e., the size of the resource pool) for performance of concurrent tasks (Lansman and Hunt 1982). If some people have more resources than others, then they should be able to perform better in dual-task circumstances than other people, but their performance loss in going from single- to dual-task conditions should be equivalent. This equivalence results because the demand for a particular task on those resources (i.e., automaticity level) will not differ. Thus, a low or zero correlation is predicted between dual-task decrement and single-task performance, and data have supported this lack of relationship (Braune and Wickens 1986; Lansman and Hunt 1982; Wickens and Weingartner 1985). Fogarty and Stankov (1982) reported that the degree of independence of the component of dual-task performance from single-task skills grows stronger as a component task is made secondary, or is deemphasized—a fact that again implicates the resource capacity metaphor to represent time-sharing ability.

These data, along with others (e.g., Fogarty and Stankov 1982, 1988; Hunt and Lansman 1981; Jennings and Chiles 1977; Stankov 1983, 1988; Sverko 1977; Wickens, Mountford, and Schreiner 1981), seem to establish that people differ in their time-sharing capabilities in ways that cannot be predicted only from their skill level on the component tasks, or automaticity. However, this conclusion has led to a further question regarding the extent to which such abilities are general, characterizing performance on a wide variety of qualitatively different dual-task pairs rather than specific to particular types of pairs (e.g., two visual tasks). It turns out that this conclusion of generality, paralleling that discussed in the context of time-sharing skill and attention switching, is extremely difficult to prove because of various statistical and methodological considerations (Ackerman, Schneider, and Wickens 1984). At best, the evidence is mixed. Wickens, Mountford, and Schreiner (1981), for example, examined differences in time-sharing ability of forty subjects performing four different tasks in nine different pairwise combinations. Although there were substantial individual differences in the efficiency of time sharing a given task pair, they did not correlate highly across the different dual-task combinations. On the other hand, Ackerman, Schneider, and Wickens (1984) examined the same data with a confirmatory factor analysis and concluded that there was some evidence for a general time-sharing ability.

Finally, given the possibility of a general time-sharing skill, it is appropriate to ask what its nature might be. Here at least three possibilities avail themselves, and they are not mutually exclusive. One is that it may relate to motivational effects, reflecting a cognitive continuum of low to high energetic arousal (Matthews and Davies 2001). High-energy people can better sustain the high resource demands of dual tasking. A second possibility is that such general ability is resident in the central executive—as discussed in chapter 9—as a means for flexibly deploying attention in demanding dual-task circumstances. This is consistent with the view of Hunt, Pellegrino, and Yee (1989) that such differences are due to attention control and coordination. A third possibility is that such differences relate to the capacity of working memory, as described following. These two components—working memory and executive control—are clearly related: Both are necessary for good multitasking, and both relate to the concept of fluid intelligence and general intelligence manifest in IQ, the single ability most predictive of success in complex multitask domains (Borman, Hanson, and Hedge 1997; Caretta and Ree 2003; Catell 1987; Hunt et al. 1983; Stankov 1987; Ree and Caretta 1996).

Attention and Working Memory

Another cognitive measure closely linked to attentional performance as well as individual differences is working memory capacity (Baddeley and Hitch

1974), a concept that can be best understood by comparison to the familiar notion of short-term memory (STM) (Atkinson and Shiffrin 1968; Miller 1956). STM, as traditionally conceived, is a cognitive system used to hold information over relatively brief periods of time. WM, in contrast, is more complex, "allowing the temporary storage and manipulation of information necessary for such complex tasks as comprehension, learning, and reasoning" (Baddeley 2000, p. 418). Thus, the WM system comprises not just the STM buffers needed to hold information but also a set of attentional control mechanisms used to regulate the flow of information into and out of STM and to transform the information stored. The influential WM model of Baddeley and Hitch (1974) and Baddeley (1986), for example, comprises a pair of short-term buffers for holding verbal and visual information, respectively, along with a central executive mechanism—as discussed in chapter 9—responsible for controlling the buffers' activity.

As befits its inclusion in the current chapter, large stable and meaningful individual differences in working memory have been well documented (Engle 2002). WM capacity can be measured with what is known as a complex span task. Various forms of complex span exist (Conway et al. 2005; Daneman and Carpenter 1980; Turner and Engle 1989), but all share the critical characteristic that they require subjects to hold information in STM while performing some additional information processing task—that is, under conditions of divided attention. In an operation span task (Turner and Engle 1989), for example, the subject is presented a series of true or false mathematical equations (e.g., $8 \times 2 - 5 = 9$), each followed by an unrelated word. The subject's task is to judge the truth of each equation while storing the words for recall at the end of the series. The task therefore demands that information be maintained in STM while additional information processing operations, which are needed to solve the equations, are performed. This can be compared with a traditional simple span task in which information is stored in STM but no additional task is performed.

Although the STM and WM systems are closely linked (Cowan 1995; Engle et al. 1999), STM and WM span scores differ dramatically in their relationship to other performance measures. STM capacity, contrary to the early expectations of many researchers, is at best weakly correlated with measures of performance on more complex tasks such as reading (Daneman and Carpenter 1980; Perfetti and Lesgold 1977). WM capacity, on the other hand, is a good predictor of many high-level cognitive abilities and traits, including reading comprehension (Daneman and Carpenter 1980), retrieval of information from long-term memory (Rosen and Engle 1997), and fluid intelligence (Engle et al. 1999). WM span scores also correlate highly with measures of various real-world skills, including proficiency in tasks such as following directions (Engle, Carullo, and Collins 1991), taking lecture notes (Kiewra and Benton 1988), learning a computer programming language (Shute 1991), and even judging the safety of a left-hand turn across lanes of oncoming traffic (Guerrier, Manivannan, and Nair 1999) (for a review see Engle 2002).

Data also show strong relationships, finally, between WM span and various lab-based measures of attentional processing. Proficiency in a selective listening task (e.g. Cherry 1953; Moray 1959; also see chapter 2 in this volume), for example, increases with WM capacity. Thus, when asked to shadow a stream of speech played to one ear and ignore a distracting voice in the other, listeners with a high WM span are less likely to notice their name in the to-be-ignored ear (Conway, Cowan, and Bunting 2001). Subjects with a high WM span also tend to suffer less Stroop interference than those with a low span (Kane and Engle 2003) and are better able to suppress capture of attention by a sudden visual transient (Kane et al. 2001). Thus, WM capacity correlates with the ability to focus attention. Furthermore, those with higher span are more flexible in distributing covert attentional resources, showing a greater ability to divide the mental spotlight of attention between noncontiguous areas of the visual field (Bleckley et al. 2003).

What makes WM span such a good predictor of performance on so many tasks, both in the lab and in real-world settings? Engle and colleagues (Engle 2002; Engle et al. 1999) argued that WM capacity predicts performance on attention-demanding tasks because the complex span task itself is a test of attentional function. As noted, the WM system includes the buffers needed to temporarily hold information and a central executive mechanism responsible for controlling the buffers' activity. Simple span tasks place little demand on the central executive mechanism because they do nothing to distract attention from information in the buffers. The complex span tasks used to measure WM capacity, however, tax the central executive heavily, requiring it to hold information in the memory buffers and protect it from interference while the focus of processing is repeatedly diverted to a concurrent task. WM capacity is a measure of how proficiently this attentional information juggling is performed; subjects with good attention control skills will better be able to maintain the information stored in short-term buffers while contending with the distracting task. WM span is a good predictor of performance in a variety of laboratory and real-world tasks, moreover, precisely because those tasks rely on the same attentional control mechanisms (Engle 2002).

Attention and Aging

Advancing age in adulthood brings a gradual but widespread decline in perceptual and cognitive performance (Fisk et al. 2004; Park 2000; Schroeder and Salthouse 2004), and it is no surprise that this includes a loss in some attentional abilities. The consequences of aging for attentional function, however, go beyond the direct effects of these losses themselves. Of perhaps equal importance is that everyday demands on attention increase with age as a result of declining sensorimotor function. As we move into middle and late adulthood, diminishing attentional resources are thus met with proliferating attentional needs. Fortunately, though, it is possible to compensate, at least in part, for these changes, through the wise application of attentional and the appropriate design of displays and tasks (Fisk and Rogers 2007; Fisk et al. 2004).

Declining Attentional Function

Notably, a number of attentional processes are spared with age. Visual search slopes for salient feature targets, for example, remain close to zero (Plude and Doussard-Roosevelt 1989), and the ability to selectively guide visual search on the basis of stimulus features remains largely or fully intact (Humphrey and Kramer 1997; Scialfa et al. 2000). Similarly, the ability to exploit both central and peripheral attentional cues in a spatial cuing task is uncompromised by healthy aging (Greenwood, Parasuraman, and Haxby 1993; Wiegmann et al. 2006), at least so long as the perceptibility of the cues themselves is matched across age groups (Gottlob and Madden 1998). Aging also appears to have little influence on processes that are highly overlearned or automatic (Hasher and Zacks 1979; Jennings and Jacoby 1993), though older adults may have difficulty achieving automatization of some new tasks (Fisk and Rogers 1991).

Other forms of attentional processing, however, are less robust across the lifespan. Some declines in attentional performance are equivalent to what is expected based on the general slowdown in perceptual and cognitive processes that occurs with advancing age (Salthouse 1996; Verhaegen 2000). As measured by differences in raw RTs, for example, older adults show greater Stroop interference and larger task-switching costs than young adults. As a proportion of baseline RTs, however, the effects are roughly the same size across young and older adult age groups (Verhaegen and Cerella 2002). In other words, advancing age seems to have no specific effects on these processes. Performance is apparently degraded only as a consequence of very broad declines in processing speed.

Other attentional mechanisms are harder hit by age, showing specific age-related losses. As one example, adults often experience particular difficulty focusing attention on task-relevant information in the presence of auditory or visual noise. Tun and Wingfield (1999), for instance, tested older and younger adults' ability to listen to and remember a stream of target speech with and without a distractor voice in the background. Memory for the target speech was compromised by the distractor for both age groups, but the effects were substantially greater for the older listeners. Age-related losses were larger still, moreover, when the distractor speech was meaningful than when it was nonmeaningful (Tun, O'Kane, and Wingfield 2002), confirming that the age difference was not simply a consequence of older listeners' sensory losses. Other data suggest similarly that older observers have difficulty processing visual information embedded among distractor items—whether those distractors are interspersed with the target spatially (e.g., Carlson et al. 1995) or temporally (Gazzaley et al. 2005)—and that the effects of visual distractors are especially pronounced when targets and distractor items do not appear in predictable locations (e.g., grouped into separate columns) (Carlson et al. 1995). These results appear, at least in part, to be the result of a deficit in top-down attentional suppression of the unwanted info (Gazzaley et al. 2005; Hasher and Zacks 1979; for caveats see Kramer et al. 1994). A related visual effect is that the visual lobe—the window of attention surrounding the point of fixation—becomes narrower with age, especially when the visual field is cluttered (Scialfa, Kline, and Lyman, 1987). The result of such attentional narrowing for older drivers is an increased risk of accident (Owsley et al. 1998). Together, these findings argue for the importance of minimizing noise levels and of ensuring clean, well-organized visual displays when designing for older users (Fisk et al. 2004).

Another form of attention strongly compromised with age is the capacity for multitasking; a large body of literature has demonstrated substantial deficits in multiple-task performance among older adults (for review see Craik 1977; Kramer, Larish, and Strayer 1995; Verhaeghen et al. 2003). In Sit and Fisk (1999), for example, young and older adult participants performed an array of four memory and perceptual judgment tasks that required them to attend to multiple visual and auditory display channels. Not surprisingly, performance for both age groups declined in the multitask condition as compared with single-task control conditions. Performance losses were disproportionately large, however, for the older adults. Similar results have been reported with task combinations ranging from walking and memorizing word lists (Li et al. 2001; Lindenberger, Marsiske, and Baltes 2000) to driving and talking on a cell phone (Alm and Nilsson 1995).

One apparent reason for the difficulty that older adults have in multitasking is that multiple-task conditions are simply more complex than single-task conditions (McDowd and Craik 1988). A second contributing factor appears to be a deficit in executive attentional control. Research has found, for instance, that older adults are less adept than their young adult counterparts in flexibly trading off attention between component tasks in a multiple-task environment (Sit and Fisk 1999; Tsang and Shaner 1998), an effect implying a loss of attentional flexibility and task coordination skills. Other findings indicate that older adults may have difficulty in simultaneously holding multiple task sets active in working memory (De Jong 2001). This conclusion is consistent with the more general finding of working memory losses in older adults (e.g., Dobbs and Rule 1989) and is bolstered by findings that in the task-switching paradigm, older adults show trial-to-trial switch costs proportionate to those of young adults but show inordinately large task-mixing costs (Bojko, Kramer, and Peterson 2004; Kray and Lindenberger 2000). Recall from Figure 9.2 that mixing costs are differences in mean RT for mixed- and pure-task blocks of trials that remain after trial-to-trial switch costs have been accounted for. They reflect the demands of keeping multiple task sets in working memory.

Increasing Attentional Demands

These reviewed findings confirm that aspects of attentional processing become less effective as we age. Evidence also indicates, unfortunately, that much of our behavior simultaneously becomes more dependent on attention. More particularly, findings show that age-related losses in sensory and physical function increase the attentional demands of everyday behavior, bringing about, in the phrasing of Lindenberger, Marsiske, and Baltes (2000, p. 434), a "permeation of behavior with cognition." Although it is common to ridicule the clumsy for their metaphorical inability to walk and chew gum at the same time, dual-task studies show that the control of posture and gait is attention demanding, particularly in the elderly (Woollacot and Shumway-Cook 2002). Thus, for older adults even the performance of relatively simple cognitive tasks—judging the pitch of a tone, detecting the onset of a light, memorizing a list of words, counting backwards by threes—can significantly compromise the ability to maintain posture and gait, to step over an obstacle, or to recover stability following a disruption of balance (e.g., Chen et al. 1996). Reciprocally, the resource demands of maintaining and recovering balance can degrade performance on secondary cognitive tasks (e.g., Lindenberger, Marsiske, and Baltes 2000). These effects are sometimes evident in the performance of young adults but are invariably larger for older adults. Why? Lindenberger et al.) speculate that as we age, "sensory and motor aspects of performance are increasingly in need of cognitive control and supervision because of frailty, sensory losses, and lower level problems in sensorimotor integration and coordination." That is, the tasks of maintaining posture and gait become more difficult as we get older, increasing their attentional requirements.

A similar phenomenon is seen at work in a connection between hearing loss and poor short-term memory. Though it is not obvious that the quality of our hearing and the capacity of our short-term memory should be related, data reveal that memory span for spoken words lists is smaller for older listeners with some hearing loss than for those with good hearing, even when the words are presented loudly enough to be heard by both groups (McCoy et al. 2005; Rabbitt 1991). The reason, theorists have concluded, is that listeners with poor hearing have to invest more effort to understand the spoken words, diverting attentional resources that could otherwise be used to keep the words active in working memory (Wingfield, Tun, and McCoy 2005). In other words, attentional resources are needed to overcome hearing losses. Consistent with this interpretation, young adults with normal hearing show lower memory spans for spoken words that are masked by noise than for words that are presented clearly (Rabbitt 1968).

Coping with Age-Related Attentional Changes

Happily, data indicate that age-related declines in cognitive performance do not translate invariably into functional everyday losses (Park and Gutchess

2000). As we age, rather, we appear to develop a variety of compensatory strategies for minimizing the consequences of our ebbing information-processing skills. One way this is accomplished in multiple task circumstances is by prioritizing the tasks that are most urgent or consequential and deemphasizing those that are less significant, optimizing the allocation of resources. Consider, for example, a study that asked young and old adults to memorize lists of words while walking a track (Li et al. 2001). As noted already, the maintenance of posture and gait is more demanding of attention in older than in younger adults. The researchers thus hypothesized that compared with single-task conditions, performance in dual-task conditions—as measured by the ability to remember the studied words and to walk without veering or stumbling—would be compromised more for older participants than for their younger counterparts. More importantly, the authors also speculated that the pattern of dual-task interference would be different across age groups. Because the consequences of stumbling and falling are worse for older than for younger adults, the authors reasoned, the older adults should be more likely to sacrifice memory performance under dual-task conditions to protect walking performance. This was exactly what the data revealed: Though dual-task costs for the walking task were similar across age groups, dual-tasks costs for the memory task were higher for older adults—in the context of chapters 7 and 8, a resource-allocation effect.

An equally important strategy for contending with age-related losses is through the use of environmental support—compensatory aids of various forms that allow operators to offload memory demands onto cues or reminders in the outside world (Baltes and Baltes 1990; Craik and Byrd 1982; Park and Gutchess 2000). The ability to exploit mechanisms of environmental support is a component of skilled performance generally and can dramatically attenuate age-related differences in cognitive performance (Fisk et al. 2004). A study by Morrow et al. (2003), for instance, compared the performance of young, middle-aged, and older pilots on a task that required participants to listen to, read back, and mentally integrate a series of simulated air-traffic commands. Subjects sometimes were allowed to take written notes while receiving their messages and sometimes were not. The results indicated that read-back accuracy declined with age only when note-taking was forbidden. When notes were allowed, accuracy was uniformly high across age groups. A follow-up study (Morrow et al. in press) found that the opportunity to take notes also reduced age-related differences in dual-task costs when operators were required to divide attention between the read-back task and piloting a flight simulator. Note-taking thus served as a form of environmental support to mitigate age-related performance losses. Interestingly, the effective use of notes as environmental support was apparently a skill that developed with experience, as note-taking failed to reduce age-related losses in a control group of nonpilot participants. Nonetheless, these results and others like them confirm that the use of environmental support can indeed help to minimize the real-world consequences of the age-related losses seen in simple laboratory measures of cognition. Li et al. (2001, p. 236) concluded, "In old age, individuals address declining abilities by prioritizing what should be preserved, and then maintaining prior performance levels by using compensatory means."

The consequences of attentional losses also can be mitigated through appropriate task design and operator training. In accordance with the predictions of multiple resource theory, research has found that age-related dual-task losses are attenuated when the paired tasks employ different response modalities (e.g., manual and vocal) rather than a single modality (e.g., manual alone) (Brouwer et al. 1991; Tsang and Shaner 1998) and, similarly, when display information is distributed over multiple input modalities. Dingus et al. (1997), for instance, demonstrated that in-vehicle displays presenting navigational information aurally rather than visually reduced demands on the visual resources needed for lane maintenance and roadway monitoring, and were especially beneficial for older drivers. These findings imply that older adults' multitasking difficulty can be reduced to the extent that processing demands are distributed across multiple attentional resource pools. The conclusion that age impairs task-set maintenance, furthermore, implies that older adults' multitask performance will be improved by environmental reminders and attentional cues that signal what tasks are to be performed and when. Indeed, the ability to exploit such mechanisms of environmental support appears to be an important component of successful aging in general (e.g., Baltes and Baltes 1990; Craik and Byrd 1982; Park and Gutchess 2000).

Finally, in accordance with the findings on attentional skill development discussed already, evidence suggests that older adults can be trained to multitask more efficiently, learning to strategically shift attention between tasks more flexibly (Sit and Fisk 1999; Tsang and Shaner 1998). Varied-priority training in particular appears to produce large gains in older adults' multitasking skills, reducing age-related differences in multiple task performance and encouraging the development of attentional control skills that generalize to novel task combinations (Kramer, Larish, and Strayer 1995; Kramer et al. 1999).

Conclusions

This chapter has considered three sources of individual differences in attentional capabilities: (1) practice, skills, and expertise; (2) traits and stable abilities; and (3) advancing age. Certainly there are other sources not treated here, including those due to clinical conditions like attention deficit disorder (Huang-Pollack and Nigg 2003); those due to young age, such as attention development across infancy and childhood (Wickens and Benel 1982); and possibly those due to cross-cultural differences. Such topics, though fascinating, are beyond the scope of the current book.

Three main themes emerge from our treatment herein. First, in understanding differences in multitask performance, it is critical to understand differences in performance of the single-task components to which attention is deployed. Second, individual differences observed in multitask performance are often largely specific to the particular task pair or task situation under consideration and are less the result of a generic skill or ability, applicable across all task combinations. This gives reason to doubt claims that the multitasking of the electronic generation will lead to substantial changes in their overall dual-tasking ability. Finally, the amalgamated combination of executive control, task management, and attention-switching capabilities—as discussed in chapter 9 and as represented so well in tests of working memory—appears to represent the most logical candidate that might underlie that which is general in attention skill and ability. The final chapter considers some of the physiological brain mechanisms that might be responsible for these and other attentional functions.

11

Cognitive Neurosciences and Neuroergonomics

Introduction

Having discussed research extending back to the work of William James (1890), it is useful to close by considering developments that portend the future of applied attention research: the convergence of human factors and neuroscience. Applied attention researchers have traditionally employed behavioral measures of performance as their primary window on mental processes. How quickly can the operator make a judgment, for example, or how much does accuracy suffer when the operator is asked to perform two tasks simultaneously? This reliance on behavioral data has been in large part for two practical reasons. First, behavioral data can usually be collected easily and with little cost, sometimes without even use of a computer. Second, performance measures such as speed and accuracy are typically the very phenomena that applied attention researchers wish to predict and control. Behavioral data are of direct and immediate interest to the topics that concern human factors scientists.

However, human factors researchers have also long held an interest in brain function (e.g., Sem-Jacobson 1959; Sem-Jacobson and Sem-Jacobson 1963). Indeed, much of the earliest research on workload was aimed at delineating the fundamental relationship between task demand and physiological measures of autonomic nervous system activity (Hockey, Coles, and Gailliard 1997; Kahneman 1973; Kalsbeek and Sykes 1967; Kramer 1987). As technological and theoretical advances have allowed cognitive neuroscience to elucidate the relationship between mental and neural processes in ever more detail, engineering psychology has gained further insights into real-world human performance. Recognizing the increasing value of brain research to applied cognitive psychology, Parasuraman (2003, p. 5) has even coined the term *neuroergonomics* to denote "the study of brain and behavior at work," the intersection of neuroscience and human factors. Of what value is neuroscience to applied psychology? Kramer and Parasuraman (2007) note possibilities. First, neuroergonomic measures may sometimes be more valid or reliable indices of psychological phenomena (e.g., effort and mental workload level) than more traditional behavioral or subjective measures. Second, neuroergonomic methods might allow measurement of psychological processes that are inaccessible to behavioral or subjective measures.

Third, neuroergonomic data might be observable under circumstances in which it is difficult or impossible to employ more traditional methods. Parasuraman (2003, p. 6) notes, moreover, that the information gained from neuroergomics can motivate, inform, and corroborate theories of cognition in applied domains, enabling researchers to "ask different questions and develop new explanatory frameworks about human and work," and can provide novel tools and guidelines for human factors practitioners. Neuroergonomics, in other words, will contribute to both the science and practice of engineering psychology.

Neuroergonomic Methods

Cognitive neuroscientists have devised a variety of methods for assaying the bodily and neural correlates of mental activity (Kramer and Parasuraman in press). At the simplest level are peripheral physiological variables such as heart rate and heart-rate variability, eye-blink frequency, and pupil diameter. Although these indices do not measure cortical activity per se, they do vary with mental task demands (Verwey and Veltman 1996) and can thus be used to infer cognitive processing load (Beatty 1982; Marshall 2007). Other measures reflect cortical activity more directly. The electroencephalogram (EEG) measures rhythmic variations in electrical activity of the brain over extended periods of time (Davidson, Jackson, and Larson 2000; Gevins and Smith 2007). The event-related potential (ERP) technique likewise measures cortical electrical activity, but time-locks the analysis to a stimulus or response event such that the time course of activity produced by perceptual, cognitive, and motor activity can be observed (Donchin, Kramer, and Wickens 1986; Fabiani, Gratton, and Coles 2000; Fu and Parasuraman 2007; Luck 2005). A pair of very different methodologies—functional magnetic resonance imaging (fMRI) (Haxby, Courtney, and Clark 1999) and positron emission tomography (PET) (Corbetta 1999)—gauge neural activation by measuring changes in blood flow that occur throughout the brain as the participant performs a cognitive task, allowing researchers to map the neural hot spots engaged by different information processing demands. Transcranial Doppler sonography (TCD) assesses neural activation levels more generally (i.e., with less spatial precision) by measuring the flow of blood into the brain (Tripp and Warm 2007). Other techniques are emerging and being refined continually (e.g., Gratton and Fabiani 2007).

These various methodologies tap different neurobiological processes, and each comes with its own benefits and constraints (Kramer and Parasuraman in press). Peripheral measures such as heart rate and pupil diameter provide only coarse information about processing load but are relatively easy and inexpensive to measure. EEG recording and analysis requires greater technical knowledge and somewhat more expensive equipment, in contrast, but provides more insight as to the operator's cognitive state. ERPs can be analyzed only if the electrical data recorded from the brain can be synchronized to external stimulus events or to the human operator's responses. However, they provide high temporal resolution and, as discussed below, can be analyzed for the presence of well-established markers of particular cognitive processes. fMRI and PET require participants to lie motionless inside a large and enormously expensive piece of equipment as data are recorded, precluding any possibility of data collection in the workplace, but they allow brain activity to be measured with high spatial detail. TCD sacrifices the spatial resolution of fMRI and PET but provides better temporal resolution, is less expensive, and allows participants greater mobility.

By itself, each of these measures provides only a circumscribed view of neural function. Together, however, these methodologies and others allow remarkably detailed knowledge of neural mechanisms and processes that underlie human cognition and performance. Data indicate that attention operates at multiple loci throughout the brain. In sensory regions of the visual cortex, limits on the number of available neurons constrain the quality of perceptual processing (Kastner and Ungerleider 2000; Luck et al. 1997). The choice of which stimuli are represented well and which are not is determined in part by perceptual salience (Beck and Kastner 2005; Reynolds and Desimone 2003) and in part by top-down control processes originating in the frontal cortex (Desimone and Duncan 1995; Hopfinger, Buonocore, and Mangun 2000). Executive control processes situated in the frontal regions of the brain also help to regulate the selection of task-relevant information, to resolve competition between conflicting responses—as in the Stroop and flanker tasks (Casey et al. 2000; Pardo et al. 1990)—and to manage task-switching operations (DiGirolamo et al. 2001).

Neuroimaging of Resources and Workload

In applied attention research, the most common use of neurocognitive measures has been for the study of mental resources and workload. A frequent criticism of resource models of attention such as those proposed by Kahneman (1973) and Wickens (1984) has been that they are overly vague, leaving it unclear exactly what mental resources are. Brain imaging data, however, have begun to identify neural correlates of the mental resources inferred from behavioral performance data. More particularly, Just, Carpenter, and Miyake (2003) argue that brain activation constitutes a form of attentional resource. Resource supply is the total available amount of brain activation, and resource demand for a given task is the amount of activation needed for task performance. Resting state brain activation levels thus provides a measure of resource availability, and activation levels during task performance provide a measure of the task's resource demands. The amount of processing that can be performed with a given resource investment depends on the efficiency of neural processing (Parks et al. 1989) as determined by factors including "neurotransmitters' functioning, the various metabolic systems supporting the neural system ... the connectivity or integrity of the neural systems" and the operator's performance strategy (Just, Carpenter, and Miyake 2003, pp. 61–62). Consistent with the cognitive theories of effort and mental load discussed in chapter 7, workload is defined as the ratio of resources demanded to resources available. Just and colleagues, echoing the earlier work of Hellige, Cox, and Litvac (1978), Polson and Friedman (1988), and Wickens and Sandry (1982), also posit separate pools of resources for spatial processing, language processing, and executive control, analogous to the multiple resource pools of Wickens' (1984, 2002) model.

Neuroimaging data comport well with this model (Just, Carpenter, and Miyake 2003). As it predicts, fMRI-measured brain activation levels increase with task difficulty. Activation levels in cortical regions responsible for verbal processing, for example, increase with sentence complexity in a language comprehension task (Just et al. 1996), and activation levels in regions involved in spatial processing increase with the difficulty of a mental rotation task (Carpenter et al. 1999). The resource demands imposed by the task, however, depend in part on the performer's ability. Thus, persons high in visual-spatial ability—as measured by an independent psychometric test—evince less cortical activation when performing a visual-spatial task than persons lower in ability, and persons high in verbal ability show less activation when performing a linguistic task than persons lower in verbal proficiency (Reichle, Carpenter, and Just 2000). When verbal and spatial tasks are performed at the same time, the summed activation levels in the language and spatial processing areas are greater than the total activation levels produced by either of the two tasks by itself but are less than the summed total of the two single tasks (Just et al. 2001), a pattern of effects analogous to that seen in behavioral performance data. The distribution of activation under dual-task conditions, moreover, can be manipulated by instructions such that activation shifts toward a task that is prioritized and away from the concurrent task (Newman, Keller, and Just 2007), again just as cognitive or attentional resources can be allocated differently in response to task instructions (as discussed in chapters 7 and 8). Data thus suggest that the verbal and spatial tasks are supported by separate networks of brain regions that draw in part from a pool of common resources that can be traded off between tasks, and in part from independent pools that can be mobilized concurrently in each processing domain. Altogether, this pattern of effects supports the notion of activation as a form of processing resource and closely parallels the premises and predictions of the multiple resource theories developed from behavioral data.

Electrophysiological Indices of Workload

Electrophysiological techniques have also proven useful for measuring and understanding mental resources and workload. EEG data show characteristic changes in spectral composition as a result of increased mental workload, with power increasing in the range of 4–8 Hz, called the theta band, and decreasing in the range of 8–12 Hz, called the alpha band (Gevins and

Smith 2007). These effects have been demonstrated in simple laboratory tasks (e.g., Gevins et al. 1998) as well as more complex and naturalistic tasks such as simulated driving (Brookhuis and de Waard 1993), flight (Sterman and Mann 1995; Sterman et al. 1994), and air-traffic control (Brookings, Wilson, and Swain 1996). Data also indicate that the theta band EEG is selectively enhanced in high performers on a working memory task under conditions of heavy cognitive load (Gevins and Smith 2000) and that theta and alpha band power both increase with practice on a task (Gevins et al. 1997). These results suggest that activity in the theta band reflects the ability to focus and sustain attention, whereas power in the alpha band is inversely correlated with the amount of neural resources recruited in task performance (Gevins and Smith 2007).

Using ERPs, attentional effects are studied by examining changes in the pattern of electrical activity that occurs in the brain following a stimulus event. The ERP waveform comprises a series of positive and negative peaks, or components, that occur at characteristic latencies and locations across the scalp (Donchin, Kramer, and Wickens 1986; Fabiani, Gratton, and Coles 2000). The influence of attention is often seen through changes in the latency or amplitude with which particular components occur. The auditory N1, for example, is a negative peak that arises from activity in the auditory cortex and begins approximately 100 ms following onset of a sound. In dichotic listening, the N1 is larger for sounds in the attended ear than for those in the unattended ear, indicating an attentional influence on very low-level auditory perception (Hillyard et al. 1973; Woldorff and Hillyard 1991). Similar effects are seen in the visual modality, where attention amplifies the P1 and N1, a sequence of positive and negative peaks that arise from the early visual cortex in response to the appearance of a stimulus (Mangun and Hillyard 1991). Thus, in both the audition and vision, attention appears to enhance very early attention selection, as discussed in chapter 2. Another component, the N2pc, has been shown to reflect the focusing of covert attention (Luck and Hillyard 1994) and can thus be used to track coarse movements of the attentional spotlight within a scene or display (Woodman and Luck 1999).

The ERP component that has been of most interest to applied attention researchers, however, has been the P3, sometimes called the P300 (Fu and Parasuraman 2007). The P3 is a positive peak with a typical latency of approximately 300 ms. It has been studied intensively, and its place within the information-processing stream of human cognition is well understood (Donchin and Coles 1988; Polich and Kok 1995; Pritchard 1981). The P3 can be elicited by auditory, visual, or tactile stimuli and is often studied by using an odd-ball paradigm in which occasional atypical stimuli are interspersed within a series of otherwise similar items (e.g., a small number of low-pitched tones within series of high-pitched tones). Regardless of the particular choice of stimuli, P3 amplitude is larger for the atypical or unexpected events than for events that are common or anticipated. This indicates that the P3 response is driven not by physical characteristics of the stimulus but by the degree to which the stimulus matches the observer's expectations (Sutton et al. 1965) and confirms that the P3 occurs after the stimulus has been recognized and evaluated against its context. Consistent with this conclusion, data show that the latency of the P3 is increased when noise masking is used to make stimulus processing more difficult. The P3 is unaffected, however, by the difficulty of response planning and execution (Kutas, McCarthy, and Donchin 1977; Magliero et al. 1984; McCarthy and Donchin 1981). The effect therefore appears to arise after stimulus evaluation but before response processing, such that P3 latency provides a measure of stimulus evaluation time. Thus, for instance, the response selection bottleneck that causes the PRP seen in response-time data has little effect on P3 latency (Luck 1998).

The P3 interests attention researchers because P3 amplitude has been shown to vary with attentional allocation. More particularly, amplitude increases as more attention is allotted to the signal that triggers the potential. The P3 can thus be used as an index of attentional processing. An experiment by Wickens et al. (1983), for example, measured P3 amplitudes to demonstrate tradeoffs in the allocation of attentional resources between auditory and visual processing in a dual task. Subjects performed a discrete visual-manual tracking task while counting auditory probes as a secondary task. P3s were recorded in response both to visual signals (i.e., movements of the target cursor) and auditory stimuli (i.e., tones to be counted), and the allocation of attention was manipulated by increasing the difficulty of the tracking task, with the expectation that attention would shift away from the auditory task as tracking became more demanding. Evidence of this shift was seen in a near-perfect trade-off between auditory and visual P3 amplitudes: As tracking difficulty rose, P3 amplitudes increased in response to visual events and decreased in response to auditory events.

Kramer, Wickens, and Donchin (1985) found similar evidence of resource reciprocity when attention was divided between two separate objects related to different tasks. Interestingly, however, no indication of resource trade-offs was found in P3 amplitudes when attention was divided between two attributes of a single object. P3 data thus lend converging evidence to the claim of Duncan's (1984) object-based theory discussed in chapter 6: that attention is allocated in parallel to all aspects of a single object. Effects of primary-task demand on P3 amplitude to secondary-task stimuli have been demonstrated using highly naturalistic tasks, including simulated air-traffic control (Isreal et al. 1980) and simulated flight with visual and instrument rules (Fowler 1991; Kramer, Sirevaag, and Braune 1987). Studies have also shown that attentional load can be measured using an irrelevant probe technique in which the operator is not required to respond to the signals that elicit the P3 (Kramer and Weber 2000). For example, P3 amplitudes in response to auditory signals are reduced by an attention-demanding task such as helicopter flight (Sirevaag et al. 1993) or radar monitoring (Kramer, Trejo, and Humphrey 1995), even when the auditory probes themselves are entirely task irrelevant. The irrelevant probe technique therefore provides a way for researchers to monitor workload levels without imposing a secondary task on the operator—though this technique may be slightly less sensitive than a manipulation that requires attention to a probe (Kramer and Weber 2000).

Operator State Assessment with Psychophysiological Measures

The examples discussed herein illustrate the value of neuroscientific techniques in developing and testing attentional theory. Of what use are these methods to the practice of applied cognitive psychology? At least two possibilities come easily to mind, both involving the assessment of operator state. First, electrophysiological measures such as the P1, N1, N2pc, and P3 may provide techniques for determining what information has been attended and encoded within a complex real-world environment. Chapter 7, for example, discussed the role of attentional effort in learning and noted that long-term memory improves when study strategies are designed to require greater effort from the learner (Schmidt and Bjork 1992). By gauging the investment of attentional resources, ERP amplitudes should be able to passively estimate how much effort a student has allotted to processing an item without the need to interrupt the learning process for tests of knowledge. Similarly, P3 amplitudes in conjunction with eye-movement data can help determine what changes in a dynamic environment are and are not noticed, thereby allowing the passive assessment of situation awareness. Here it is noteworthy that although eye fixations cannot reveal inattentional blindness—the looked but did not see effect—the P3 is able to do so. Strayer and Drews (2007), for example, examined the effects of cell phone conversations on eye movements and ERPs in a simulated driving task. ERPs were elicited by an occasional critical event: onset of the brake lights of a car in front of the participant. Data revealed that the number of fixations on the critical events was similar under single- and dual-task conditions but that cell phone conversations reduced P3 amplitude. Participants looked at the critical events in both conditions, in other words, but attended to them less when carrying on a conversation.

Second, neuroergonomic techniques can be used to monitor mental workload in near real time. As noted, various psychophysiological measures (e.g., EEG patterns, ERP amplitudes, heart rate, pupil diameter) vary in response to workload increases, and data suggest that such variables can in fact be used to estimate workload levels with high reliability over intervals as brief as 10 to 30 seconds (Gevins et al. 1998; Humphrey and Kramer 1994; Verwey and Veltman 1996). These measures have the benefit that they can typically be collected unobtrusively without the need to impose a secondary task or otherwise interrupt task performance. They are likely to be particularly helpful, moreover, as an increasing reliance on automation raises the ratio of an operator's unobservable cognitive activity to observable physical activity, making inferences of workload from behavioral measures more difficult and less reliable.

An emerging use of physiological workload monitoring is in the adaptive control of automated systems (Rouse 1988; Schmorrow et al. 2006). A danger in implementing automated aids is that the automation may sometimes provide too little assistance and sometimes too much. In the former case, human workload levels may become excessively high when task conditions become demanding. In the latter case, the human operator may become complacent toward his or her task, neglecting to monitor the automated aid's performance or the raw data on which the aid is operating (Parasuraman, Molloy, and Singh 1993) and may experience work underload, a decrease in arousal leading to poor task performance (Young and Stanton 2002). A way to avoid these problems is to adapt the level of automation in response to the changes in the operator's workload level, providing more assistance when workload is high and less when workload levels are low (Hillburn et al. 1997; Parasuraman et al. 1996; Scerbo 2001).

The ability to adapt an automated aid to the operator's workload, however, requires some technique for monitoring workload levels as the operator interacts with the system. Researchers have suggested that electrophysiological methods are well suited for this purpose (Byrne and Parasuraman 1996; Gomer 1981), and data have supported this speculation (Pope, Bogart, and Bartolome 1995; Scerbo, Freeman, and Mikulka 2003). An experiment by Prinzel et al. (2000), for example, asked participants to perform a simulated flight task battery involving manual tracking, gauge monitoring, and fuel management. Under some conditions, the manual-tracking component of the battery was automated. For one group of participants, the assignment of the manual-tracking task to either the operator or the automation was adaptively controlled by an EEG-based monitoring system that switched responsibility for tracking to the automation when operator workload was high and back to the operator when workload decreased. For the remaining participants, switches between automated and operator control of the tracking task took place at preset times independent of the operator's workload. The authors reasoned that if the system was effective at detecting high workload levels, then tracking performance should have been better and subjective workload should have been lower for participants aided by the EEG-based system than for those in the control group. Data confirmed this prediction (Prinzel et al. 2000; see also Freeman et al. 1999, 2000). A follow-up study found that assistance from the EEG-based adaptive automation also reduced operators' complacency and improved their situation awareness in a flight simulation task (Bailey et al. 2003). Altogether, these results suggest that neuroergonomic techniques will play a valuable role in the design of future human–machine systems.

Conclusions

We have covered a great amount of ground since chapter 1, from James's (1890) early observations on attention, dating back well over a century, to the explosion in behavioral attention research over the past fifty years, to the burgeoning understanding of brain mechanisms in attention, to the future applications of this understanding in harmonizing human–system interaction. Despite the tremendous amount of knowledge that has accrued through this long span of research, there is little evidence that the attentional demands we contend with at home and in the workplace are diminishing or will ever do so. As the influence of mechanical and electronic automation seeps further into our lives, rather, the world simply claims less from our motor resources and more from our cognition, less from our muscles and more from our minds. The attentional demands we face are changing, not dissipating. Only by sharing knowledge between the lab and the world— drawing applications from theory and motivating theoretical discovery with the promise of new application—will we understand and meet these evolving attentional challenges.

References

Ackerman, P., Schneider, W., and Wickens, C. D. (1984). Deciding the existence of a time-sharing ability: A combined methodological and theoretical approach. *Human Factors* 26: 71–82.

Adams, M. J., Tenny, Y., and Pew, R. W. (1991). *Strategic workload and the cognitive management of multi-task systems*. CSERIAC SOAR. Wright Patterson AFB Ohio: Crew System Ergonomics Information Analysis Center.

Aldrich, T. B., Szabo, S. M., and Bierbaum, C. R. (1989). The development and application of models to predict operator workload during system design. In G. R. McMillan, D. Beevis, E. Salas, M. H. Strub, R. Sutton, and L. Van Breda (Eds.), *Applications of human performance models to system design* (pp. 65–80). New York: Plenum.

Alexander, A., Wickens, C.D. and Hardy, T. J. (2005). Synthetic vision systems: The effects of guidance symbology, display size, and field of view. *Human Factors, 47,* 693–707.

Allport, A. D., Antonis, B., and Reynolds, P. (1972). On the division of attention: A disproof of the single channel hypothesis. *Quarterly Journal of Experimental Psychology,* 24: 225–235.

Allport, A., Styles, E. A., and Hsieh, S. (1994). Shifting intentional set: Exploring the dynamic control of tasks. In Umiltà and Moscovitch, pp. 451–452.

Alm, H. and Nilsson, L. (1995). The effects of a mobile telephone task on driver behaviour in a car following situation. *Accident Analysis and Prevention* 27(5): 707–715.

Altmann, G. and Trafton, J. (2002). Memory for goals. *Cognitive Science* 23: 39–83.

Amlôt, R., Walker, R., Spence, C., and Driver, J. (2003). Multimodal visual-somatosensory integration in saccade generation. *Neuropsychologia* 41(1): 1–15.

Andre, A. D. and Wickens, C. D. (1992). Layout analysis for cockpit display systems. SID International Symposium Digest of Technical Papers. Paper presented at the Society for Information Display, Seattle, WA, June.

Aretz, A. J. (1991). The design of electronic map displays. *Human Factors* 33(1): 85–101.

Atchley, P. and Dressel, J. (2004). Conversation limits the functional field of view. *Human Factors* 46: 664–673.

Atkinson, R. C. and Shiffrin, R. M. (1968). Human memory: A proposed system and its control processes. In K. W. Spence and J. T. Spence (Eds.), *The psychology of learning and motivation*, vol. 2 (pp. 89–195). New York: Academic Press.

Baddeley, A. (1996). Exploring the central executive. *Quarterly Journal of Experimental Psychology* 49A: 5–28.

Baddeley, A. D. (1986). *Working memory*. Oxford: Clarendon Press.

Baddeley, A. D. (2000). The episodic buffer: A new component of working memory. *Trends in Cognitive Science* 4: 417–423.

Baddeley, A. D., Chincotta, D., and Adlam, A. (2001). Working memory and the control of action: Evidence from task switching. *Journal of Experimental Psychology: General* 130: 641–657.

Baddeley, A. D. and Hitch, G. J. (1974). Working memory. In G. Bower (Ed.), *Recent advances in learning and motivation*, vol. 8. New York: Academic Press.

Bahrick, H. P., Noble, M., and Fitts, P. M. (1954). Extra task performance as a measure of learning a primary task. *Journal of Experimental Psychology* 48: 298–302.

Bahrick, H. P. and Shelley, C. (1958). Time Sharing as an Index of Automization. *Journal of Experimental Psychology* 56: 288–293.

Bailey, B. P. and Konstan, J. A. (2006). On the need for attention-aware systems: Measuring effects of interruption on task performance, error rate, and affective state. *Computers in Human Behavior* 23: 685–708.

Bailey, N. R., Scerbo, M. W., Freeman, F. G., Mikulka, P. J., and Scott, L. A. (2003). The effects of a brain-based adaptive automation system on situation awareness: The role of complacency potential. In *Proceedings of the Human Factors and Ergonomics Society 47th annual meeting* (pp. 1048–1052). Santa Monica, CA: Human Factors and Ergonomics Society.

Bainbridge, L. (1983). Ironies of automation. *Automatica* 19(6): 775–779.

Ball, K. K., Beard, B. L., Roenker, D. L., Miller, R. L., and Griggs, D. S. (1988). Age and visual search: Expanding the useful field of view. *Journal of Optical Society of America* A(4): 2210–2279.

Ballard, D. H., Hayhoe, M. M., and Pelz, J. B. (1995). Memory representation in natural tasks. *Journal of Cognitive Neuroscience* 7(1): 66–86.

Baltes, P. B. and Baltes, M. M. (1990). Psychological perspectives on successful aging: The model of selective optimization with compensation. In P. B. Baltes and M. M. Baltes (Eds.), *Successful aging: Perspectives from the behavioral sciences* (pp. 1–34). New York: Cambridge University Press.

Barclay, R. L., Vicari, J. J., Doughty, A. S., Johanson, J. F., and Greenlaw, R. L. (2006). Colonoscopic withdrawal times and adenoma detection during screening colonoscopy. *New England Journal of Medicine* 355: 2533–2541.

Barnett, B. J. and Wickens, C. D. (1988). Display proximity in multicue information integration: The benefit of boxes. *Human Factors* 30: 15–24.

Bartram, L., Ware, C., and Calvert, T. (2003). Moticons: Detection, distraction, and task. *International Journal of Human-Computer Studies* 58: 515–545.

Beatty, J. (1982). Task-evoked puillary responses, processing load, and the structure of processing resources. *Psychological Bulletin* 91: 276–292.

Beatty, J. and Kahneman, D. (1966). Pupilllary changes in two memory tasks. *Psychonomic Science* 5: 371–372.

Beck, D. M. and Kastner, S. (2005). Stimulus context modulates competition in human extrastriate cortex. *Nature Neuroscience* 8: 1110–1116.

Bellenkes, A. (1999). The use of pilot performance models to facilitate cockpit visual scanning. Unpublished dissertation, University of Illinois.

Bellenkes, A., Wickens, C. D., and Kramer, A. (1997). Visual scanning and pilot expertise. *Aviation Space and Environmental Medicine* 48(7): 569–579.

Bennett, K. B. and Flach, J. M. (1992). Graphical displays: Implications for divided attention, focused attention, and problem solving. *Human Factors* 34: 513–533.

Bennett, K. B., Toms, M. L., and Woods, D. D. (1993). Emergent features and graphical elements: Designing more effective configural displays. *Human Factors* 35(1): 71–98.

Beringer, D. B. and Chrisman, S. E. (1991). Peripheral polar-graphic displays for signal/failure detection. *International Journal of Aviation Psychology* 1: 133–148.

Bertelson, P. (1966). Central intermittency twenty years later. *Quarterly Journal of Experimental Psychology* 18: 153–163.

Bettman, J. R., Johnson, E. J., and Payne, J. (1990). A componential analysis of cognitive effort and choice. *Organizational Behavior and Human Performance* 45: 111–139.

Bettman, J. R., Payne, J. W., and Staelin, R. (1986). Cognitive considerations in designing effective labels for presenting risk information. *Journal of Marketing and Public Policy* 5: 1–28.

Bialystok, E. (1999). Cognitive complexity and attentional control in the bilingual mind. *Child Development* 70: 636–644.

Biederman, I., Glass, A. L., and Stacy, E. W. J. (1973). Searching for objects in real-world scenes. *Journal of Experimental Psychology* 97: 22–27.

Bjork, R. A. (1999). Assessing our own competence: Heuristics and illusions. In Gopher A. and Koriat (Eds.). *Attention and Performance XVII*, Cambridge, MA: MIT Press, pp. 435–459.

Bleckley, M. K., Durso, F. T., Crutchfield, J. M., Engle, R. W., and Khana, M. M. (2003). Individual differences in working memory capacity predict visual attention allocation. *Psychonomic Bulletin and Review* 10: 884–889.

Bliss, J. (2003). An investigation of alarm related accidents and incidents in aviation. *International Journal of Aviation Psychology* 13(3): 249–268.

Boff, K. R. (Ed.), Kaufman, L., and Thomas, J. P. (1986). *Handbook of perception and human performance: Cognitive processes and performance.* New York: Wiley and Sons.

Bojko, A., Kramer, A. F., and Peterson, M. S. (2004). Age equivalence in switch costs for prosaccade and antisaccade tasks. *Psychology and Aging* 19: 226–234.

Boot, W. R., Kramer, A. F., Becic, E., Wiegmann, D. A., and Kubose, T. (2006). Detecting transient changes in dynamic displays: The more you look, the less you see. *Human Factors* 48: 759–773.

Borman, W. C., Hanson, M. A., and Hedge, J. W. (1997). Personnel selection. *Annual Review of Psychology* 48: 299–337.

Brandimonte, M., Einstein, G., and McDaniel, M. (1996). *Prospective memory: Theory and applications.* Hillsdale, NJ: Erlbaum.

Braune, R. and Wickens, C. D. (1986). Time-sharing revisited: Test of a componential model for the assessment of individual differences. *Ergonomics* 29(11): 1399–1414.

Breznitz, S. (1983). *Cry-wolf: The psychology of false alarms.* Hillsdale, NJ: Lawrence Earlbaum.

Broadbent, D. (1958). *Perception and communications.* New York: Permagon.

Broadbent, D. (1982). Task combination and selective intake of information. *Acta Psychologica* 50: 253–290.

Brookhuis, K. A. and de Waard, D. (1993). The use of psychophysiology to assess driver status. *Ergonomics* 36: 1099–1110.

Brookings, J. and Damos, D. (1991). Individual differences in multiple task performance. In Damos (Ed.). *Multiple Task Performance*, London: Taylor & Francis, pp. 363–386.

Brookings, J., Wilson, G. F., and Swain, C. R. (1996). Psychophysiological responses to changes in workload during simulated air traffic control. *Biological Psychology* 42: 361–377.

Brouwer, W. H., Waterink, W., Van Wolffelaar, P. C., and Rothengatter, T. (1991). Divided attention in experienced young and older drivers: Lane tracking and visual analysis in a dynamic driving simulator. *Human Factors* 33: 573–582.

Burns, C. M. (2000). Putting it all together: Improving integration in ecological displays. *Human Factors* 42: 226–241.

Burns, C. M. and Hajdukiewicz, J. R. (2004). *Ecological interface design.* Boca Raton, FL: CRC Press.

Bursill, A. E. (1958). The restriction of peripheral vision during exposure to hot and humid conditions. *Quarterly Journal of Experimental Psychology* 10: 113–129.

Byrne, E. A. and Parasuraman, R. (1996). Psychophysiology and adaptive automation. *Biological Psychology* 42: 249–268.

Byrne, M. D., Anderson, J. R., Douglass, S., and Matessa, M. (1999). Eye tracking the visual search of click-down menus. *Human Factors in Computing Systems: Proceedings of CHI 99*, pp. 402–409.

Cacioppo, J. T., Tassinary, L. G., and Berntson, G. G. (Eds.) (2000). *Handbook of psychophysiology*, 2d ed. Cambridge, UK: Cambridge University Press.

Cades, D. M., Trafton, G., T., and Boehm-Davis, D. (2005). Mitigating disruptions: Can resuming an interrupted task be trained? Paper presented at the *Proceedings of the 50th Annual Conference of the Human Factors and Ergonomics Society*, Santa Monica, CA.

Carbonnell, J. F., Ward, J., and Senders, J. (1968). A queuing model of visual sampling: Experimental validation. *IEEE Transactions on Man Machine Systems* MMS-9: 82–72.

Caretta, T. and Ree, M. J. (2003). Pilot selection methods. In M. Vidulich and P. Tsang (Eds.), *Principles and Practices of Aviation Psychology* (pp. 357–396). Mahwah, NJ: Lawrence Erlbaum.

Carlson, M. C., Hasher, L., Connelly, L. S., and Zacks, R. T. (1995). Aging, distraction, and the benefits of predictable location. *Psychology and Aging* 10: 427–436.

Carpenter, P. A., Just, M. A., Keller, T. A., Eddy, W. F., and Thulborn, K. R. (1999). Graded functional activation in the visuospatial system with the amount of task demand. *Journal of Cognitive Neuroscience* 11: 9–24.

Carpenter, S. (2002). Sights unseen. *APA Monitor* 32: 54–57.

Carrasco, M., Penpeci-Talgar, C., and Eckstein, M. (2000). Spatial covert attention increases contrast sensitivity across the CSF: Support for signal enhancement. *Vision Research* 40: 1203–1215.

Carroll, J. M. and Carrithers, C. (1984). Blocking learner error states in a training-heels system. *Human Factors* 26: 377–389.

Carswell, C. M. and Wickens, C. D. (1996). Mixing and matching lower-level codes for object displays: Evidence for two sources of proximity compatibility. *Human Factors* 38(1): 1–22.

Casey, B. J., Thomas, K. M., Welsh, T. F., Badgaiyan, R. D., Eccard, C. H., Jennings, J. R., et al. (2000). Dissociation of response conflict, attentional selection, and expectancy with functional magnetic resonance imaging. *Proceedings of the National Academy of Sciences* 97(15): 8728–8733.

Cattel, R. (1987). *Intelligence: Its structure, growth and action*. New York: Elsevier.

Cave, K. R. and Bichot, N. P. (1999). Visuospatial attention: Beyond a spotlight model. *Psychonomic Bulletin and Review* 6(2): 204–223.

Cellier, J. and Eyrolle, H. (1992). Interference between switched tasks. *Ergonomics* 35(1): 25–36.

Chan, A. H. S. and Courtney, A. J. (1996). Foveal acuity, peripheral acuity, and search performance: A review. *International Journal of Industrial Ergonomics* 18: 113–119.

Chen, H. C., Schultz, A. B., Ashton-Miller, J. A., Giordani, B., Alexander, N. B. and Guire, K. E. (1996). Stepping over obstacles: Dividing attention impairs performance of old more than young adults. *Journals of Gerontology Series A: Biological Sciences and Medical Sciences, 51*, 116–122.

Chernikoff, R., Duey, J. L., and Taylor, F. V. (1960). Two dimensional tracking with identical and different control dynamics in each coordinate. *Journal of Experimental Psychology* 60: 318–322.

Cherry, C. (1953). Some experiments on the recognition of speech with one and with two ears. *Journal of the Acoustical Society of America* 25: 975–979.

Chi, M.T.H. (2000). Self-explaining expository texts: The dual processes of generating inferences and repairing mental models. In R. Glaser (Ed.). *Advances in Instructional Design*.(pp161–238). Mahwah, NJ: Erlbaum.

Chou, C., Madhavan, D., and Funk, K. (1996). Studies of cockpit task management errors. *International Journal of Aviation Psychology* 6(4): 307–320.

Chun, M. M. and Jiang, Y. (1998). Contextual cuing: Implicit learning and memory of visual context guides spatial attention. *Cognitive Psychology, 36*, 28–71.

Chun, M. M. and Jiang, Y. (1999). Top-down attentional guidance based on implicit learning of visual covariation. *Psychological Science* 10: 360–365.

Chun, M. M. and Nakayama, K. (2000). On the functional role of implicit visual memory for the adaptive deployment of attention across scenes. *Visual Cognition* 7: 65–81.

Chun, M. M. and Wolfe, J. M. (1996). Just say no: How are visual searches terminated when there is no target present? *Cognitive Psychology* 30: 39–78.

Cole, W. G. (1986). Medical cognitive graphics. *Proceedings of the Association for Computing Machinery Inc.* ACM-SIGCHI Human Factors in Computing Systems. New York.

Conway, A. R. A., Cowan, N., and Bunting, M. F. (2001). The cocktail party phenomenon revisited: The importance of working memory capacity. *Psychonomic Bulletin and Review* 8(2): 331–335.

Conway, A. R. A., Kane, M. J., Bunting, M. F., Hambrick, D. Z., Wilhelm, O., and Engle, R. W. (2005). Working memory span tasks: A methodological review and user's guide. *Psychonomic Bulletin and Review* 12: 769–786.

Corbetta, M. (1998). Frontoparietal cortical networks for directing attention and the eye to visual locations: Identical, independent, or overlapping neural systems? *Proceedings of the National Academy of Sciences* 95: 831–838.

Corbetta, M. (1999). Functional anatomy of visual attention in the human brain: Studies with positron emission tomography. In R. Parasuraman (Ed.), *The attentive brain* (pp. 95–122). Cambridge, MA: MIT Press.

Cousineau, D. and Shiffrin, R. M. (2004). Termination of a visual search with large display size effects. *Spatial Vision* 17: 327–352.

Cowan, N. (1995). *Attention and memory: An integrated framework.* Oxford: Oxford University Press.

Craik, F. I. M. (1977). Age differences in human memory. In J. E. Birren and K. W. Schaie (Eds.), *Handbook of the psychology of aging* (pp. 384–420). New York: Van Rostrand Reinhold.

Craik, F. I. M. and Byrd, M. (1982). Aging and cognitive deficits: The role of attentional resources. In F. I. M. Craik and S. Trehub (Eds.), *Aging and cognitive processes* (pp. 191–211). New York: Plenum Press.

Craik, F. I. M. and Lockhart, R. S. (1972). Levels of processing: L a framework for memory research. *Journal of Verbal Learning and Verbal Behavior* 11: 671–684.

Craik, K. W. J. (1947). Theory of the human operator in control systems I: The operator as an engineering system. *British Journal of Psychology* 38: 56–61.

Damos, D. (1978). Residual attention as a predictor of pilot performance. *Human Factors* 20: 435–440.

Damos, D. (Ed.) (1991). *Multiple task performance.* London: Taylor and Francis.

Damos, D., Smist, T., and Bittner, A. (1983). Individual differences in multiple task performance as a function of response strategies. *Human Factors* 25: 215–226.

Damos, D. and Wickens, C. D. (1980). The identification and transfer of time-sharing skills. *Acta Psychologica* 46: 15–39.

Damos, D. L. (1997). Using interruptions to identify task prioritization in Part 121 air carrier operations. In R. Jensen (Ed). *Proceedings of the 9th International Symposium on Aviation Psychology,* Ohio State University, Columbus.

Daneman, M. and Carpenter, P. A. (1980). Individual differences in working memory and reading. *Journal of Verbal Learning and Verbal Behavior* 19: 450–466.

Davidson, R. J., Jackson, D. C., and Larson, C. L. (2000). Human electroencephalography. In Cacioppo, Tassinary, and Berntson, pp. 27–52.

Davis, R. (1965). Expectancy and intermittency. *Quarterly Journal of Experimental Psychology* 17: 75–78.

De Jong, R. (2001). Adult age differences in goal activation and goal maintenance. *European Journal of Cognitive Psychology* 13: 71–89.

Degani, A. and Wiener, E. L. (1993). Cockpit checklists: Concepts, design and use. *Human Factors* 35: 345–360.

Desimone, R. and Duncan, J. (1995). Neural mechanisms of selective visual attention. *Annual Review of Neuroscience* 18: 193–222.

Deubel, H. and Schneider, W. X. (1996). Saccade target selection and object recognition: Evidence for a common attentional mechanism. *Vision Research* 36: 1827–1837.

Deutch, J. and Deutch, D. (1963). Attention: Some theoretical considerations. *Psychological Review* 70: 80–90.

DiGirolamo, G. J., Kramer, A. F., Barad, V., Cepeda, N. J., Weissman, D. H., Milham, M. P., et al. (2001). General and task-specific frontal lobe recruitment in older adults during executive processes: A fMRI investigation of task-switching. *Neuroreport* 12: 2065–2072.

Dingus, T. A., Hulse, M. C., Mollenhauer, M., Fleischman, R. N., McGehee, D. V., and Manakkal, N. (1997). Effects of age, system experience, and navigation technique on driving with an advanced traveler information system. *Human Factors* 39: 177–199.

Dingus, T. A., Klauer, S. G., Neale, V. L., Petersen, A., Lee, S. E., Sudweeks, J., et al. (2006). The 100-car naturalistic driving study, phase II—Results of the 100-car field experiment. Technical Report DOT HS 810 593. Washington, DC: National Highway Traffic Safety Administration.

Dismukes, K. (2001). The challenge of managing interruptions, distractions, and deferred tasks. In R. Jensen (Ed.). *Proceedings of the 11th International Symposium on Aviation Psychology*, Ohio State University, Columbus, CD-Rom.

Dismukes, K. and Nowinski, J. (2007). Prospective memory, concurrent task management and pilot error. In A. Kramer, D. Wiegmann, and A. Kirlik, *Attention: From theory to practice*, Oxford: Oxford University Press, pp. 225–238.

Dixon, S. R. and Wickens, C. D. (2006). Automation reliability in unmanned aerial vehicle control: A reliance-compliance model of automation dependence. *Human Factors* 48: 474–468.

Dixon, S. R., Wickens, C. D., and Chang, D. (2005). Mission control of multiple UAVs: A quantitative workload analysis. *Human Factors* 47: 479–487.

Dixon, S. R., Wickens, C. D., and McCarley, J. (2007). On the independence of reliance and compliance: Are automation false alarms worse than misses. *Human Factors* 49, pp. 564–572.

Dobbs, A. R. and Rule, B. G. (1989). Adult age differences in working memory. *Psychology and Aging* 4: 500–503.

Dodhia, R. and Dismukes, R. K. (2003). *A task interrupted becomes a prospective memory task*. Wellington, New Zealand: Society for Applied Research in Memory and Cognition.

Donchin, E. and Coles, M. G. H. (1988). Is the P300 component a manifestation of context updating? *Behavioural Brain Sciences* 11: 357–374.

Donchin, E., Kramer, A. F., and Wickens, C. D. (1986). Applications of brain event-related potentials to problems in engineering psychology. In M. G. H. Coles, E. Donchin, and S. W. Porges (Eds.), *Psychophysiology: Systems, processes, and applications* (pp. 702–718). New York: Guilford Press.

Dornheim, M. A. (2000, July 17). Crew distractions emerge as new safety focus. *Aviation Week and Space Technology*, 58–65.

Downing, C. J. and Pinker, S. (1985). The spatial structure of visual attention. In M. I. Posner and O. S. M. Marin (Eds.), *Attention and performance,* vol. 11 (pp. 171–187). Hillsdale, NJ: Erlbaum.

Drews, F. and Strayer, D. (2007). Multi-tasking in the automobile. In A. Kramer, D. Wiegmann, and A. Kirlik (Eds.). *Attention: From theory to practice*, Oxford University Press, pp. 121–133.

Driver, J. and Spence, C. (1994). Spatial synergies between auditory and visual attention. In C. Umiltà and M. Moscovitch (Eds.), *Attention and performance, vol. 15: Conscious and nonconscious information processing* (pp. 311–331). Cambridge, MA: MIT Press.

Driver, J. and Spence, C. (Eds.) (2004). *Crossmodal space and crossmodal attention.* Oxford: Oxford University Press.

Drury, C. G. (1975). Inspection of sheet metals—Model and data. *Human Factors* 17: 257–265.

Drury, C. G. (1990). Visual search in industrial inspection. In D. Brogan (Ed.), *Visual search* (pp. 263–276). London: Taylor and Francis.

Drury, C. G. and Chi, C. F. (1995). A test of economic models of stopping policy in visual search. *IIE Transactions* 27: 382–393.

Drury, C. G. and Wang, M.-J. (1986). Are research results in inspection task specific? *Proceedings of the Human Factors Society 30th annual meeting*, Santa Monica, CA, 476–479.

Duncan, J. (1979). Divided attention: The whole is more than the sum of its parts. *Journal of Experimental Psychology: Human Perception and Performance* 5: 216–228.

Duncan, J. (1984). Selective attention and the organization of visual information. *Journal of Experimental Psychology: General* 119: 501–517.

Duncan, J. and Humphreys, G. W. (1989). Visual search and stimulus similarity. *Psychological Review* 96: 433–458.

Dupont, V. and Bestgen, Y. (2006). Learning from technical documents: The role of intermodal referring expressions. *Human Factors* 8: 257–264.

Durso, F., Nickerson, R. S., Dumais, S. T., Lewandowsky, S., and Perfect, T. J. (Eds.) (2007). *Handbook of applied cognition,* 2d ed. Chichester, UK: John Wiley.

Durso, F., Rawson, K., and Girotto, S. (2007). Comprehension and situation awareness. In F. Durso (Ed.). *Handbook of Applied Cognition*, 2nd ed. Chichester UK: John Wiley & Sons, pp. 163–194.

Easterbrook, J. A. (1959). The effect of emotion on cue utilization and the organization of behavior. *Psychological Review* 66: 183–201.

Edworthy, J., Loxley, S., and Dennis, I. (1991). Improved auditory warning design: Relations between warning sound parameters and perceived urgency. *Human Factors* 33: 205–231.

Egeth, H. E., Virzi, R. A., and Garbart, H. (1984). Searching for conjunctively defined targets. *Journal of Experimental Psychology: Human Perception and Performance* 10: 32–39.

Ellis, S. R. and Stark, L. (1986). Statistical dependency in visual scanning. *Human Factors* 28(4): 421–438.

Endsley, M. R. (1995). Toward a theory of situation awareness in dynamic systems. *Human Factors* 37: 85–104.

Endlsey, M. R. (2006). Situation awareness. In G. Salvendy (Ed.). *Handbook of Human Factors and Ergonomics*, 3rd ed. pp.528–542.

Endsley, M. R. and Kiris, E. O. (1995). The out of the loop performance problem and level of control in automation. *Human Factors* 37(2): 381–394.

Engle, R. W. (2002). Working memory capacity as executive attention. *Current Directions in Psychological Science* 11: 19–23.

Engle, R. W., Carullo, J. J., and Collins, K. W. (1991). Individual differences in working memory for comprehension and following directions. *Journal of Educational Research* 84: 253–262.

Engle, R. W., Tuholski, S. W., Laughlin, J. E., and Conway, A. R. A. (1999). Working memory, short-term memory, and general fluid intelligence: A latent-variable approach. *Journal of Experimental Psychology: General* 128: 309–331.

Ericsson, K. A. and Simon, H. A. (1993). *Protocol analysis.* Cambridge, MA: MIT Press.

Eriksen, B. A. and Eriksen, C. W. (1974). Effects of noise letters upon the identification of a target letter in a nonsearch task. *Perception and Psychophysics* 16: 143–149.

Eriksen, C. W. and Hoffman, J. E. (1972). Temporal and spatial characteristics of selective encoding from visual displays. *Perception and Psychophysics* 12: 201–204.

Eriksen, C. W. and Hoffman, J. E. (1973). The extent of processing of noise elements during selective encoding from visual displays. *Perception and Psychophysics* 14(1): 155–160.

Eriksen, C. W. and St. James, J. D. (1986). Visual attention within and around the field of focal attention: A zoom lens model. *Perception and Psychophysics* 40: 225–240.

Fabiani, M., Buckley, J., Gratton, G., Coles, M. G., Donchin, E., and Logie, R. (1989). The training of complex task performance. *Acta Psychologica* 71: 259–299.

Fabiani, M., Gratton, G., and Coles, M. G. H. (2000). Event-related brain potentials: Methods, theory, and applications. In J. Cacioppo, L. Tassinary, and G. Berntson, *Handbook of Psychophysiology*, 3rd ed., Cambridge University Press, pp. 53–84.

Fadden, S., Ververs, P., and Wickens, C. D. (1998). Costs and benefits of head-up display use: A meta-analytic approach. *Proceedings of the 42nd Annual Meeting of the Human Factors and Ergonomics Society*, Santa Monica, CA.

Fadden, S., Ververs, P. M., and Wickens, C. D. (2001). Pathway HUDS: Are they viable? *Human Factors* 43(2): 173–193.

Fagot, C. and Pashler, H. (1992). Making two responses to a single object: Exploring the central bottleneck. *Journal of Experimental Psychology: Human Perception and Performance* 18: 1058–1079.

Fennema, M. G. and Kleinmuntz, D. N. (1995). Anticipations of effort and accuracy in multiattribute choice. *Organizational Behavior and Human Decision Processes* 63(1): 21–32.

Fenton, J. J., Taplin, S. H., Carney, P. A., Abraham, L., Sickles, E. A., D'Orsi, C., et al. (2007). Influence of computer-aided detection on performance of screening mammography. *New England Journal of Medicine* 356: 1399–1409.

Fisher, D. and Pollatsek, A. (2007). Novice Driver Crashes: Failure to divide attention or failure to recognize risks. In A. Kramer, D. Wiegmann, and A. Kirlik, *Attention: From Theory to Practice*, Oxford University Press, pp. 134–156.

Fisher, D. L., Coury, B. G., Tengs, T. O., and Duffy, S. A. (1989). Minimizing the time to search visual displays: The role of highlighting. *Human Factors* 31(2): 167–182.

Fisher, D. L. and Tan, K. C. (1989). Visual displays: The highlighting paradox. *Human Factors* 31: 17–31.

Fisk, A. D., Ackerman, P. L., and Schneider, W. (1987). Automatic and controlled processing theory and its applications to human factors. In P. A. Hancock (Ed.), *Human factors psychology* (pp. 159–198). Amsterdam: North-Holland.

Fisk, A. D. and Rogers, W. A. (1991). Toward an understanding of age-related memory and visual search effects. *Journal of Experimental Psychology: General* 120: 131–149.

Fisk, A. D. and Rogers, W. (2007). Attention goes home: Support for aging adults. In A. Kramer, D. Wiegmann, and A. Kirlik, *Attention: From Theory to Practice*, Oxford University Press, pp. 157–169.

Fisk, A. D., Rogers, W. A., Charness, N., Czaja, S. J., and Sharit, J. (2004). *Designing for older adults*. Boca Raton, FL: CRC Press.

Fisk, A. and Schneider, W. (1981). Controlled and automatic processing during tasks requiring sustained attention. *Human Factors* 23: 737–750.

Fisk, A.D. and Schneider, W. (1983). Category and word search: Generalizing search principles to complex processing. *Journal of Experimental Psychology: Learning Memory and Cognition* 9: 117–195.

Fitts, P., Jones, R. E., and Milton, E. (1950). Eye movements of aircraft pilots during instrument landing approaches. *Aeronautical Engineering Review* 9: 24–29.

Fleetwood, M. D. and Byrne, M. D. (2006). Modeling the visual search of displays: A revised ACT-R model of icon search based on eye-tracking data. *Human-Computer Interaction* 21: 153–198.

Fogarty, G. and Stankov, L. (1982). Competing tasks as an index of intelligence. *Personality and Individual Differences* 3: 407–722.

Fogarty, G. and Stankov, L. (1988). Abilities involved in performance on competing tasks. *Personality and Individual Differences* 9: 35–50.

Folk, C. L., Remington, R. W., and Johnston, J. C. (1992). Involuntary covert orienting is contingent on attentional control settings. *Journal of Experimental Psychology: Human Perception and Performance* 18: 1030–1044.

Foster, D. H. and Ward, P. A. (1991). Asymmetries in oriented-line detection indicate two orthogonal filters in early vision. *Proceedings of the Royal Society of London, B, 243,* 75–81.

Fowler, F.D. (1980). Air Traffic control problems: a pilot's view. *Human Factors, 22,* 645–654.

Foyle, D. and Hooey, B. (Eds.) (2007). *Pilot performance models.* Mahwah NJ: Lawrence Erlbaum.

Fracker, M. L. and Wickens, C. D. (1989). Resources, confusions, and compatibility in dual axis tracking: Displays, controls, and dynamics. *Journal of Experimental Psychology: Human Perception and Performance* 15: 80–96.

Freed, M. (2000). Reactive prioritization. Paper presented at the *Proceedings of the 2nd NASA International Workshop on Planning and Scheduling in Space,* San Francisco, CA.

Freeman, F. G., Mikulka, P. J., Prinzel, L. J., and Scerbo, M. W. (1999). Evaluation of an adaptive automation system using three EEG indices with a visual tracking task. *Biological Psychology* 50: 61–76.

Freeman, F. G., Mikulka, P. J., Scerbo, M. W., Prinzel, L. J., and Clouatre, K. (2000). Evaluation of a psychophysiologically controlled adaptive automation system using performance on a tracking task. *Applied Psychophysiology and Biofeedback* 25: 103–115.

Frens, M. A. and Van Opstal, A. J. (1995). A quantitative study of auditory-evoked saccadic eye movements in two dimensions. *Experimental Brain Research* 107: 103–117.

Fu, S. and Parasuraman, R. (2007). Event-related potentials (ERPs) in neuroergonomics. In Parasuraman and Rizzo, *Neuroergonomics.* Oxford University Press, pp. 32–50.

Funk, K. H. (1991). Cockpit task management: Preliminary definitions, normative theory, error taxonomy, and design recommendations. *International Journal of Aviation Psychology* 1(4): 271–285.

Gale, N., Golledge, R. G., and Pellegrino, J. W. (1990). The acquisition and integration of route knowledge in an unfamiliar neighborhood. *Journal of Environmental Psychology* 10: 3–25.

Gale, N., Golledge, R., Pellegrino, J. W., and Doherty, S. (1990). The acquisition and integration of route knowledge in an unfamiliar environment. *Journal of Experimental Psychology* 10, 3–25.

Garner, W. R. (1974). *The processing of information and structure*. Hillsdale, NJ: Erlbaum.

Gazzaley, A., Cooney, J. W., Rissman, J., and D'Esposito, M. (2005). Top-down suppression deficit underlies working memory impairment in normal aging. *Nature Neuroscience* 8: 1298–1300.

Getty, D. J., Swets, J. A., Pickett, R. M., and Gonthier, D. (1995). System operator response to warnings of danger: A laboratory investigation of the effects of the predictive value of a warning on human response time. *Journal of Experimental Psychology: Applied* 1(1): 19–33.

Gevins, A. and Smith, M. E. (2000). Neurophysiological measures of working memory and individual differences in cognitive ability and cognitive style. *Cerebral Cortex, 10*, 829–839.

Gevins, A. and Smith, M. E. (2007). Electroencephalography (EEG) in neuroergonomics. In Parasuraman and Rizzo, pp. 15–31.

Gevins, A., Smith, M. E., Leong, H., McEvoy, L., Whitfield, S., Du, R., et al. (1998). Monitoring working memory load during computer-based tasks with EEG pattern recognition methods. *Human Factors* 40: 79–91.

Gevins, A., Smith, M. E., McEvoy, L., and Yu, D. (1997). High-resolution EEG mapping of cortical activation related to working memory: Effects of task difficulty, type of processing, and practice. *Cerebral Cortex* 7: 374–385.

Gibson, B. S. and Kingstone, A. (2006). Semantics of space: Beyond central and peripheral cues. *Psychological Science* 17: 622–627.

Gibson, J. J. (1979). The ecological approach to visual perception. Hillsdale, NJ: Erlbaum.

Gigorenzer, G. and Todd, P (1999). *Simple heuristics that make us smart*. New York: Oxford University Press.

Gillan, D. J., Wickens, C. D., Hollands, J. G., and Carswell, C. M. (1998). Guidelines for presenting quantitative data in HFES publications. *Human Factors* 40(1): 28–41.

Gillie, T. and Broadbent, D. (1989). What makes interruptions disruptive? A study of length, similarity, and complexity. *Psychological Research* 50: 243–250.

Goettl, B. P., Wickens, C. D., and Kramer, A. F. (1991). Integrated displays and the perception of graphical data. *Ergonomics* 34: 1047–1063.

Gomer, F. E. (1981). Physiological monitoring and the concept of adaptive systems. In J. Morel and K. F. Kraiss (Eds.), *Manned systems design* (pp. 271–287). New York: Plenum.

Gonzales, V. G. and Mark, G. (2004). Constant multi-tasking craziness: Managing multiple working spheres. In *Human Factors of Computing Systems: CHI 04* (pp. 113–120). New York: ACM.

Gopher, D. (1982). A selective attention test as a prediction of success in flight training. *Human Factors* 24: 173–184.

Gopher, D. (1992). The skill of attention control: Acquisition and execution of attention strategies. In S. Kornblum and D. Meyer (Eds.), *Attention and performance XIV* (pp. 299–322). Cambridge, MA: MIT Press.

Gopher, D. (1996). Attention control: Explorations of the work of an executive controller. *Cognitive Brain Research* 5: 23–38.

Gopher, D. (2007). Emphasis change in high demand task training. In A. Kramer, D. Wiegmann, and A. Kirlik, *Attention: From Theory to Practice*, Oxford University Press, pp. 209–224.

Gopher, D., Armony, L., and Greenshpan, Y. (2000). Switching tasks and attention policies. *Journal of Experimental Psychology: General* 129: 308–339.

Gopher, D. and Brickner, M. (1980). On the training of time-sharing skills: An attention viewpoint. *Proceedings of the 24th Annual Meeting of the Human Factors Society*, Santa Monica, CA.

Gopher, D. and Kahneman, D. (1971). Individual difference in attention and the prediction of flight criteria. *Perceptual and Motor Skills* 33: 1335–1342.

Gopher, D. and Koriat, A. (Eds.) (1999). *Attention and performance,* vol. 17. Cambridge, MA: MIT Press.

Gopher, D. and Sanders, A. (1985). S-Oh-R: Oh stages oh resources! In W. Prinz and A. F. Sanders (Eds.), *Cognition and motor processes* (pp. 231–253). Amsterdam: North Holland.

Gopher, D., Weil, M., and Barakeit, T. (1994). Transfer of skill from a computer game trainer to flight. *Human Factors* 36(4): 387–405.

Gopher, D., Weil, M., and Siegel, D. (1989). Practice under changing priorities: An approach to training of complex skills. *Acta Psychologica* 71: 147–197.

Gottlob, L. R. and Madden, D. J. (1998). Time course of allocation of visual attention after equating for sensory differences: An age-related perspective. *Psychology and Aging* 13(1): 138–149.

Gramopadhye, A. K., Drury, C. G., Jiang, X., and Sreenivasan, R. (2002). Visual search and visual lobe size: Can training one affect the other? *International Journal of Industrial Ergonomics* 30: 181–195.

Gramopadhye, A. K., Drury, C. O., and Prabhu, P. (1997). Training for aircraft visual inspection. *Human Factors and Ergonomics in Manufacturing* 3: 171–186.

Gratton, G. and Fabiani, M. (2007). Optical imaging of brain function. In Parasuraman and Rizzo, pp. 65–81.

Gray, W. (2000). The nature and processing of errors in interactive behavior. *Cognitive Science* 24: 205–248.

Gray, W. (2007). *Integrated models of cognitive systems.* Boca Raton, FL: Taylor and Francis.

Gray, W., Sims, C., Fu, W.-T., and Schoelles, M. J. (2006). The soft constraints hypothesis: A rational analysis approach to resource allocation for interactive behavior. *Psychological Review* 113: 461–482.

Gray, W. D. and Fu, W. T. (2004). Soft constraints in interactive behavior: The case of ignoring perfect knowledge in-the-world for imperfect knowledge in-the-head. *Cognitive Science* 28: 359–382.

Green, B. F. and Anderson, L. K. (1956). Color coding in a visual search task. *Journal of Experimental Psychology* 51: 19–24.

Green, C. S. and Bavelier, D. (2003). Action video game modifies visual selective attention. *Nature* 423: 534–537.

Green, D. M. and Swets, J. A. (1966). *Signal detection theory and psychophysics.* Oxford: John Wiley.

Greenwood, P. M., Parasuraman, R., and Haxby, J. V. (1993). Changes in visuospatial attention over the adult lifespan. *Neuropsychologia* 31(5): 471–485.

Grimes, J. (1996). On the failure to detect changes in scenes across saccades. In K. Akins (Ed.), *Perception: Volume 2. Vancouver Studies in Cognitive Science* (pp. 89–110). New York: Oxford University Press.

Guerrier, J. H., Manivannan, P., and Nair, S. M. (1999). The role of working memory, field dependence, visual search, and reaction time in the left turn performance of older female drivers. *Applied Ergonomics* 30: 109–119.

Harris, J. R. and Wilkins, A. (1982). Remembering to do things: A theoretical framework and an illustrative experiment. *Human Learning* 1: 123–136.

Hart, S. G. (1989). Crew workload-management strategies: A critical factor in system performance. In R. Jensen (Ed.). *Proceedings of the Fifth International Symposium on Aviation Psychology* Columbus, OH: Ohio State University, Department of Aviation.

Hart, S. G. and Staveland, L. E. (1988). Development of NASA-TLX (Task Load Index): Results of empirical and theoretical research. In P. A. Hancock and N. Meshkati (Eds.), *Human mental workload* (pp. 139–183). Amsterdam: North Holland.

Hart, S. G. and Wickens, C. D. (1990). Workload assessment and prediction. In H. R. Booher (Ed.), *MANPRINT: An approach to systems integration* (pp. 257–296). New York: Van Nostrand Reinhold.

Hasher, L. and Zacks, R. (1979). Automatic and effortful processes in memory. *Journal of Experimental Psychology: General* 108: 356–388.

Haskell, I. D. and Wickens, C. D. (1993). Two- and three-dimensional displays for aviation: A theoretical and empirical comparison. *International Journal of Aviation Psychology* 3(2): 87–109.

Haxby, J. V., Courtney, S. M., and Clark, V. P. (1999). Functional magnetic resonance imaging and the study of attention. In R. Parasuraman (Ed.), *The attentive brain* (pp. 123–142). Cambridge, MA: MIT Press.

Helleberg, J. and Wickens, C. D. (2003). Effects of data-link modality and display redundancy on pilot performance: An attentional perspective. *International Journal of Aviation Psychology* 13(3): 189–210.

Hellige, J., Cox, P., and Litvac, L. (1978). Formation processing in the cerebral hemispheres: Selective hemispheric activation and capacity limitations. *Journal of Experimental Psychology: General* 108: 251–279.

Helmholtz, H. (1925). *Helmholtz's treatise on physiological optics, translated from the 3rd German edition* (Vol. III). Rochester, NY: The Optical Society of America.

Hendy, K. C., Liao, J., and Milgram, P. (1997). Combining time and intensity effects in assessing operator information-processing load. *Human Factors* 39: 30–47.

Herrmann, D., Brubaker, B., Yoder, C., Sheets, V., and Tio, A. (1999). Devices that remind. In Durso et al., pp. 377–407.

Herslund, M. B. and Jørgensen, N. O. (2003). Looked-but-failed-to-see-errors in traffic. *Accident Analysis and Prevention* 35: 885–891.

Hess, S. M. and Detweiller, M. (1994). Training to reduce the disruptive effects of interruptions. *Proceedings of the 38th Conference of the Human Factors Society*, Santa Monica, CA, 1173–1177.

Hilburn, B., Jornal, P.G., Byrne, E.A., and Parasuraman, R. (1997). The effects of adaptive air traffic control (ATC) decision aiding on controller mental workload. In M. Mouloua, and J. Koonce (Eds). *Human Automation Interaction: Research and Practice.* (pp 84–91). Mahwah, NJ: Erlbaum.

Hill, S. G., Iavecchia, H., Byers, J., Bittner, A., Zaklad, A., and Christ, R. (1992). Comparison of four subjective workload rating scales. *Human Factors* 34: 429–440.

Hillyard, S. A., Hink, R. F., Schwent, V. L., and Picton, T. W. (1973). Electrical signs of selective attention in the human brain. *Science* 182: 177–179.

Hirst, W. (1986). Aspects of divided and selected attention. In J. LeDoux and W. Hirst (Eds.), *Mind and brain.* New York: Cambridge University Press.

Hirst, W. and Kalmar, D. (1987). Characterizing attentional resources. *Journal of Experimental Psychology: General* 116: 68–81.

Ho, C. and Spence, C. (2005). Assessing the effectiveness of various auditory cues in capturing a driver's visual attention. *Journal of Experimental Psychology: Applied* 11: 157–174.

Ho, C., Tan, H. Z., and Spence, C. (2005). Using spatial vibrotactile cues to direct visual attention in driving scenes. *Transportation Research Part F* 8: 397–412.

Ho, C. Y., Nikolic, M. I., Waters, M., and Sarter, N. B. (2004). Not now! Supporting interruption management by indicating the modality and urgency of pending tasks. *Human Factors* 46: 399–410.

Ho, G., Scialfa, C. T., Caird, J. K., and Graw, T. (2001). Visual search for traffic signs: The effects of clutter, luminance, and aging. *Human Factors* 43: 194–207.

Hochberg, J. and Brooks, V. (1978). Film cutting and visua momentum. In J. Senders, D. Fisher, and R. Monty (Eds.), *Eye movements and the higher psychological functions*. Hillsdale, NJ: Erlbaum.

Hockey, A. G., Gailliard, A., and Coles, M. E. (1986). *Energetics and human information processing*. Dordrecht, Netherlands: Martinus Nijhoff.

Hockey, G. R. (1997). Compensatory control in the regulation of human performance under high stress and workload. *Biological Psychology* 45: 73–93.

Hockey, R., Gaillard, A. and Coles, M. (Eds.), *Energetics and human information processing*. Dordrecht, Netherlands: Martinus Nijhoff.

Hoffman, J. E. and Subramaniam, B. (1995). The role of visual attention in saccadic eye movements. *Perception and Psychophysics* 57: 787–795.

Holnagel, E. (2007). Human error: Trick or treat? In F. T. Durso et al. (Ed.). *Handbook of Applied Cognition*, 2nd ed. New York: Wiley, pp. 219–234.

Hopfinger, J. B., Buonocore, M. H., and Mangun, G. R. (2000). The neural mechanisms of top-down attentional control. *Nature Neuroscience* 3: 284–291.

Hornof, A. J. (2001). Visual search and mouse-pointing in labeled versus unlabeled two-dimensional visual hierarchies. *ACM Transactions on Computer-Human Interaction* 8: 171–197.

Horowitz, T. S. and Wolfe, J. M. (1998). Visual search has no memory. *Nature* 394: 575–577.

Horrey, W.J. and Wickens, C.D. (2003). Multiple resource modeling of task interference in vehicle control, hazard awareness and in-vehicle task performance. *Proceedings of the Second International Driving Symposium on Human Factors in Driver Assessment, Training, and Vehicle Design, Park City, Utah*, 7–12.

Horrey, W. J. and Wickens, C. D. (2004a). Driving and side task performance: The effects of display clutter, separation, and modality. *Human Factors* 46(4): 611–624.

Horrey, W. J. and Wickens, C. D. (2004b). Focal and ambient visual contributions and driver visual scanning in lane keeping and hazard detection. *Proceedings of the 48th Annual Meeting of the Human Factors and Ergonomics Society*, Santa Monica, CA.

Horrey, W. J. and Wickens, C. D. (2006). The impact of cell phone conversations on driving: A meta-analytic approach. *Human Factors* 48(1): 196–205.

Horrey, W. J. and Wickens, C. D. (2007). In-vehicle glance duration: Distributions, tails and a model of crash risk. Transportation Research Record Annual Meeting Paper.

Horrey, W. J., Wickens, C. D., and Consalus, K. P. (2006). Modeling drivers' visual attention allocation while interacting with in-vehicle technologies. *Journal of Experimental Psychology: Applied* 12(2): 67–78.

Hoyer, W. J. and Ingolfsdottir, D. (2003). Age, skill and contextual cuing in target detection. *Psychology and Aging* 18: 210–218.

Huang-Pollock, C. L. and Nigg, J. T. (2003). Searching for the attention deficit in attention deficit hyperactivity disorder. *Clinical Psychology Review* 604, 801–830.

Humphrey, D. G. and Kramer, A. F. (1994). Toward a psychophysiological assessment of dynamic changes in mental workload. *Human Factors, 36*, 3–26.

Humphrey, D. G. and Kramer, A. F. (1997). Age differences in visual search for feature, conjunction, and triple-conjunction targets. *Psychology and Aging* 12, 704–717.

Hunt, E. and Lansman, M. (1981). Individual differences in attention. In R. J. Sternberg (Ed.), *Advances in the psychology of human intelligence,* vol. 1. Hillsdale, NJ: Erlbaum.

Hunt, E., Pellegrino, J. W., and Yee, P. L. (1989). Individual differences in attention. *Psychology of Learning and Motivation* 24: 285–310.

Hunter, J. E. and Burke, R. F. (1995). *Handbook of pilot selection.* Aldershot, UK: Averbury Aviation.

Iani, C. and Wickens, C. D. (2007). Factors affecting task management in aviation. *Human Factors* 49(1): 16–24.

Ikeda, M. and Takeuchi, T. (1975). Influence of foveal load on the functional visual field. *Perception and Psychophysics* 18: 255–260.

Irwin, D. E., Colcombe, A. M., Kramer, A. F., and Hahn, S. (2000). Attentional and oculomotor capture by onset, luminance, and color singletons. *Vision Research* 40: 1443–1458.

Isreal, J., Chesney, G., Wickens, C. D., and Donchin, E. (1980). P300 and tracking difficulty: Evidence for a multiple capacity view of attention. *Psychophysiology* 17: 259–273.

Isreal, J., Wickens, C. D., Chesney, G., and Donchin, E. (1980). The event-related brain potential as an index of display-monitoring workload. *Human Factors* 22: 211–224.

Itti, L. and Koch, C. (2000). A saliency-based search mechanism for overt and covert shifts of visual attention. *Vision Research* 40: 1489–1506.

Jacobs, A. M. (1986). Eye-movement control in visual search: How direct is visual span control? *Perception and Psychophysics* 39: 47–58.

James, W. (1890). *Principles of Psychology.* New York: Holt.

Jennings, A. E. and Chiles, W. D. (1977). An investigation of time-sharing ability as a factor in complex task performance. *Human Factors* 19: 535–547.

Jennings, J. M. and Jacoby, L. L. (1993). Automatic versus intentional uses of memory: Aging, attention, and control. *Psychology and Aging, 8,* 283–293.

Jennings, J. M. and Jacoby, L. L. (1997). An opposition procedure for detecting age-related deficits in recollection: Telling effects of repetition. *Psychology and Aging* 12: 352–361.

Jersild, A. T. (1927). Mental set and shift. *Archives of Psychology, Whole 89.*

Johnson, E. J. and Payne, J. W. (1985). Effort and accuracy in choice. *Management Science* 31: 395–414.

Johnson, E. J., Payne, J. W., and Bettman, J. R. (1988). Information displays and preference reversals. *Organizational Behavior and Human Decision Processes* 42: 1–21.

Jolicoeur, P. and Ingleton, M. (1991). Size invariance in curve tracing. *Memory and Cognition* 19(1): 21–36.

Jonides, J. (1980). Towards a model of the mind's eyes' movement. *Canadian Journal of Psychology* 34: 103–112.

Jonides, J. (1981). Voluntary versus automatic control over the mind's eye's movement. In J. Long and A. Baddeley (Eds.), *Attention and performance IX* (pp. 187–203). Hillsdale, NJ: Erlbaum.

Jonides, J. and Yantis, S. (1988). Uniqueness of abrupt visual onset as an attention–capturing property. *Perception and Psychophysics, 43*, 346–354.

Just, M., Carpenter, P. A., and Miyake, A. (2003). Neuroindices of cognitive workload: Neuroimaging, pupillometric and event-related brain potential studies of brain work. *Theoretical Issues in Ergonomics Science* 4: 56–88.

Just, M. A., Carpenter, P. A., Keller, T. A., Eddy, W. F., and Thulborn, K. R. (1996). Brain activation modulated by sentence comprehension. *Science* 274: 114–116.

Just, M. A., Carpenter, P. A., Keller, T. A., Emery, L., Zajac, H., and Thulborn, K. R. (2001). Interdependence of nonoverlapping cortical systems in dual cognitive tasks. *Neuroimage* 14: 417–426.

Kahneman, D. (1973). *Attention and effort.* Englewood Cliffs, NJ: Prentice Hall.

Kahneman, D. (2003). A perspective on judgment and choice: mapping bounded rationality. *American Psychologist. 58*, 607–720.

Kahneman, D., Ben-Ishae, R., and Lotan, M. (1973). Relation of a test of attention to road accidents. *Journal of Applied Psychology* 58: 113–115.

Kahneman, D., Slovic, P., and Tversky, A. (Eds.) (1982). *Judgment under uncertainty: Heuristics and biases.* New York: Cambridge University Press.

Kahneman, D. and Treisman, A. (1984). Changing views of attention and automaticity. In R. Parasuraman and R. Davies (Eds.). *Varieties of Attention*, Orlando, FL: Academic Press. pp. 29–61.

Kalsbeek, J.W. and Sykes, R.W. (1967). Objective measurement of mental load. *Acta Psychologica, 27,* 253–261.

Kane, M. J., Bleckley, M. K., Conway, A. R. A., and Engle, R. W. (2001). A controlled-attention view of WM capacity. *Journal of Experimental Psychology: General* 130: 169–183.

Kane, M. J. and Engle, R. W. (2003). Working-memory capacity and the control of attention: The contributions of goal neglect, response competition, and task set to Stroop interference. *Journal of Experimental Psychology: General* 132: 47–70.

Kantowitz, B. H. (1974). Double stimulation. In B. H. Kantowitz (Ed.), *Human information processing* (pp. 83–132). Hillsdale NJ: Erlbaum.

Kantowitz, B. H. and Knight, J. L. (1976). Testing tapping timesharing: I. Auditory secondary task. *Acta Psychologica* 40: 343–362.

Karlin, L. and Kestinbaum, R. (1968). Effects of number of alternatives on the psychological refractory period. *Quarterly Journal of Experimental Psychology* 20: 160–178.

Kastner, S. and Ungerleider, L. G. (2000). Mechanisms of visual attention in the human cortex. *Annual Review of Neuroscience* 23: 315–341.

Keele, S. W. (1973). *Attention and human performance.* Pacific Palisades, CA: Goodyear.

Kiewra, K. A. and Benton, S. L. (1988). The relationship between information processing ability and notetaking. *Contemporary Educational Psychology* 13: 33–44.

Kirwan, B. and Ainsworth, L. (1992). *A guide to task-analysis.* London: Taylor and Francis.

Klapp, S. T. (1979). Doing two things at once: The role of temporal compatibility. *Memory and Cognition* 9: 398–401.

Klein, G.S. (1964). Semantic power peasured through the interference of words with color naming. *American Journal of Psychology, 77,* 576–588.

Koch, C. and Ullman, S. (1985). Shifts in visual attention: Towards the underlying circuitry. *Human Neurobiology* 4: 219–222.

Kosslyn, S. M. (1994). *The elements of graph design*. New York: Freeman and Co.

Kowler, E., Anderson, E., Dosher, B., and Blaser, E. (1995). The role of attention in the programming of saccades. *Vision Research* 35: 1897–1916.

Kraiss, K. F. and Knäeuper, A. (1982). Using visual lobe area measurements to predict visual search performance. *Human Factors* 24: 673–682.

Kramer, A. F. and Hahn, S. (1995). Splitting the beam: Distribution of attention over noncontiguous regions of the visual field. *Psychological Science* 6(6): 381–386.

Kramer, A. F., Humphrey, D. G., Larish, J. F., Logan, G. D., and Strayer, D. L. (1994). Aging and inhibition: Beyond a unitary view of inhibitory processing in action. *Psychology and Aging* 9(4): 491–512.

Kramer, A. F., Larish, J. F., and Strayer, D. L. (1995). Training for attentional control in dual-task settings: A comparison of young and old adults. *Journal of Experimental Psychology: Applied* 1(1): 50–76.

Kramer, A. F., Larish, J. L., Weber, T. A., and Bardell, L. (1999). Training for executive control: Task coordination strategies and aging. In D. Gopher and A. Koriat, *Attention and Performance XVII*, Cambridge, MA: MIT Press, pp. 617–652.

Kramer, A. F. and Parasuraman, R. (2007). Neuroergonomics: Application of neuroscience to human factors. In J. T. Cacioppo, L. G. Tassinary and G. G. Berntson (Eds.), *Handbook of psychophysiology* (pp. 704–722). Cambridge, UK: Cambridge University Press.

Kramer, A. F., Sirevaag, E., and Braune, R. (1987). A psychophysiological assessment of operator workload during simulated flight missions. *Human Factors* 29: 145–160.

Kramer, A. F., Trejo, L., and Humphrey, D. G. (1995). Assessment of mental workload with task-irrelevant auditory probes. *Biological Psychology* 40: 83–100.

Kramer, A. F. and Weber, T. (2000). Applications of psychophysiology to human factors. In J. T. Cacioppo, L. G. Tassinary, and G. G. Berntson (Eds.), *Handbook of psychophysiology*, 3d ed. (pp. 794–814). Cambridge, UK: Cambridge University Press.

Kramer, A. F., Wickens, C. D., and Donchin, E. (1985). Processing of stimulus properties: evidence for dual-task integrality. *Journal of Experimental Psychology: Human Perception and Performance* 11(4): 393–408.

Kramer, A., Wiegmann, D., and Kirlik, A. (Eds.) (2007). *Attention: From theory to practice*. Oxford: Oxford University Press.

Kray, J. and Lindenberger, U. (2000). Adult age differences in task switching. *Psychology and Aging* 15: 126–147.

Kristofferson, A. B. (1967). Successiveness discrimination as a two-state quantal process. *Science* 158: 1337–1340.

Kroft, P. D. and Wickens, C. D. (2003). Displaying multi-domain graphical database information: An evaluation of scanning, clutter, display size, and user interactivity. *Information Design Journal* 11(1): 44–52.

Krois, P. (1999). Human factors assessment of the URET conflict probe false alert rate. Paper presented at the *Federal Aviation Administration*.

Kundel, H. L. and LaFollette, P. S. (1972). Visual search patterns and experience with radiological images. *Radiology* 103: 523–528.

Kundel, H. L. and Nodine, C. F. (1975). Interpreting chest radiographs without visual search. *Radiology* 116: 527–532.

Kundel, H. L., Nodine, C. F., and Carmody, D. (1978). Visual scanning, pattern recognition, and decision-making in pulmonary nodule detection *Investigative Radiology* 13: 175–181.

Kutas, M., McCarthy, G., and Donchin, E. (1977). Augmenting mental chronometry: The P300 as a measure of stimulus evaluation time. *Science* 197: 792–795.

LaBerge, D. (1973). Attention and the measurement of perceptual learning. *Memory and Cognition* 1: 268–276.

LaBerge, D. and Brown, V. (1989). Theory of attentional operations in shape identification. *Psychological Review, 96*, 101–124.

Lachter, J., Forster, K. I., and Ruthruff, E. (2004). Forty-five years after Broadbent (1958): Still no identification without attention. *Psychological Review*, 111, 880–913.

Lansman, M. and Hunt, E. (1982). Individual differences in secondary task performance. *Memory and Cognition* 10, 10–24.

Lansman, M., Poltrock, S. E., and Hunt, E. (1983). Individual differences in the ability to focus and divide attention. *Intelligence* 12: 299–312.

Lappin, J. (1967). Attention in the identification of stimuli in complex visual displays. *Journal of Experimental Psychology* 75: 321–328.

Latorella, K. A. (1996). Investigating interruptions—An example from the flightdeck. In *Proceedings of the 40th Annual Meeting HFES*. Santa Monica, CA: Human Factors.

Laudeman, I. V. and Palmer, E. A. (1995). Quantitative measurement of observed workload in the analysis of aircrew performance. *International Journal of Aviation Psychology* 5(2): 187–198.

Laughery, K. R., LeBiere, C., and Archer, S. (2006). Modeling human performance in complex systems. In G. Salvendy, *Handbook of human factors and ergonomics*, 3rd ed. Wiley, pp. 967–996.

Leachtenauer, J. C. (1978). Peripheral acuity and photointerpretation performance. *Human Factors* 20: 537–551.

Leahy, W. and Sweller, J. (2004). Cognitive load theory and the imagination effect. *Applied Cognitive Psychology* 18: 857–875.

Leahy, W. and Sweller, J. (2005). Interactions among the imagination, expertise reversal and element interactivity effects. *Journal of Experimental Psychology: Applied* 11(4): 266–276.

Lee, E. and MacGregor, J. (1985). Minimizing users each time in menu retrieval systems. *Human Factors* 27: 157–162.

Lee, J. and Moray, N. (1994). Trust, self-confidence and operator's adaptation to automation. *International Journal of Human Computer Studies* 40: 153–184.

Lee, J. D. and See, K. A. (2004). Trust in technology: Designing for appropriate reliance. *Human Factors* 46(1): 50–80.

Leibowitz, H. W., Post, R. B., Brandt, T., and Dichgans, J. W. (1982). Implications of recent developments in dynamic spatial orientation and visual resolution or vehicle guidance. In W. Wertheim and Leibowitz (Eds.), *Tutorials on motion perception* (pp. 231–260). New York: Plenum Press.

Levin, D. T., Momen, N., Drivdahl, S. B., and Simons, D. J. (2000). Change blindness blindness: The metacognitive error of overestimating change-detection ability. *Visual Cognition* 7: 397–412.

Levine, M. (1982). You-are-here-maps: Psychological considerations. *Environment and Behavior* 14: 221–237.

Levy, J., Pashler, H. and Boer, E. (2006). Central interference in driving: Is there any stopping the psychological refractory period? *Psychological Science*, 17, 228–235.

Lewandowski, S., Little, D., and Kalish, M. L. (2007). Knowledge and expertise. In Durso et al., pp. 83–110.

Li, K. Z., Lindenberger, U., Freund, A. M., and Baltes, P. B. (2001). Walking while memorizing: Age-related differences in compensatory behavior. *Psychological Science* 12: 230–237.

Liao, J. and Moray, N. (1993). A simulation study of human performance deterioration and mental workload. *Le Travail Humain* 56(4): 321–344.

Lindenberger, U., Marsiske, M., and Baltes, P. B. (2000). Memorizing while walking: Increase in dual-task costs from young to adulthood to old age. *Psychology and Aging* 15: 417–436.

Lintern, G. and Wickens, C. D. (1991). Issues for acquisition in transfer of timesharing and dual-task skills. In Damos, pp. 123–138.

Liu, Y. and Wickens, C. D. (1992a). Visual scanning with or without spatial uncertainty and divided and selective attention. *Acta Psychologica* 79: 131–153.

Liu, Y. and Wickens, C. D. (1992b). Use of computer graphics and cluster analysis in aiding relational judgment. *Human Factors* 34(2): 165–178.

Lloyd, D. M., Merat, N., McGlone, F., and Spence, C. (2003). Crossmodal links between audition and touch in covert endogenous spatial attention. *Perception and Psychophysics* 65: 901–924.

Logan, G. (1988). Toward an instance theory of automatization. *Psychological Review* 95: 492–527.

Logan, G. (2002). An instance theory of attention and memory. *Psychological Review* 109: 376–400.

Logan, G. D. and Gordon, R. D. (2001). Executive control of visual attention in dual-task situations. *Psychological Review, 108*, 393–434.

Logan, G., Kramer, A., and Coles, M. (1996). *Converging operations in the study of attention*. Washington, DC: APA Press.

Lohr, S. (2007). *Slow down brave multitasker, and don't read this in traffic*. New York: New York Times, CVLI.

Lowe, R. K. (2003). Animation and learning: Selective processing information in dynamic graphics. *Learning and Instruction* 13: 157–176.

Lu, Z. L. and Dosher, B. (1998). External noise distinguishes attention mechanisms. *Vision Research* 38: 1183–1198.

Luck, S. J. (1998). Sources of dual-task interference. *Psychological Science* 9: 223–227.

Luck, S. J. (2005). *An introduction to the event-related potential technique*. Cambridge, MA: MIT Press.

Luck, S. J., Girelli, M., McDermott, M. T., and Ford, M. A. (1997). Bridging the gap between monkey neurophysiology and human perception: An ambiguity resolution theory of visual selective attention. *Cognitive Psychology* 33: 64–87.

Luck, S. J. and Hillyard, S. A. (1994). Electrophysiological correlates of feature analysis during visual search. *Psychophysiology, 31*, 291–308.

Mac Gregor, D., Fischoff, B., and Blackshaw, L. (1987). Search success and expectations with a computer interface. *Information Processing Management* 23: 419–432.

Mack, A. and Rock, I. (1998). *Inattentional blindness*. Cambridge, MA: MIT Press.

Mackworth, N. H. (1965). Visual noise causes tunnel vision. *Psychonomic Science* 3: 67–68.

MacLeod, C. M. (1992). The Stroop task: The "gold standard" of attentional measures. *Journal of Experimental Psychology: General* 121: 12–13.

Madhavan, P., Weigmann, D., and Lacson, F. (2006). Automation failures on tasks easily performed by operators undermine trust in automated aids. *Human Factors* 48: 241–256.

Magliero, A., Bashore, T., Coles, M. G. H., and Donchin, E. (1984). On the dependence of P300 latency on stimulus evaluation processes. *Psychophysiology* 21: 171–186.

Maltz, M. and Shinar, D. (2003). New alternative methods in analyzing human behavior in cued target acquisition. *Human Factors* 45: 281–295.

Maltz, M. and Shinar, D. (2004). Imperfect in-vehicle collision avoidance warning systems can aid drivers. *Human Factors* 46: 357–366.

Mangun, G. R. and Hillyard, S. A. (1991). Modulations of sensory-evoked brain potentials indicate changes in perceptual processing during visual-spatial priming. *Journal of Experimental Psychology: Human Perception and Performance* 17: 1057–1074.

Marshall, S.P (2007). Identifying cognitive state from eye metrics. *Aviation Space and Environmental Medicine 78*, 5, B165–B174.

Marshall, D. C., Lee, J. D., and Austria, P. A. (2007). Alerts for in-vehicle information systems: Annoyance, urgency and appropriateness. *Human Factors* 49: 145–157.

Martin, G. (1989). The utility of speech input in user-computer interfaces. *International Journal of Man-Machine System Study* 18: 355–376.

Martin, R. C., Wogalter, M. S., and Forlano, J. G. (1988). Reading comprehension in the presence of unattended speech and music. *Journal of Memory and Language* 27: 382–398.

Matthews, G. and Davies, D. R. (2001). Individual differences in arousal and sustained attention: A dual task study. *Personality and Individual Differences* 31: 571–589.

Mayer, R. E. (1999). Instructional technology. In F. Durso (Ed.), *Handbook of applied cognition 1st ed.*, (pp. 551–570). Chichester, UK: Wiley Interscience.

Mayer, R. E., Hegarty, M., Mayer, S., and Campbell, J. (2005). When static media promote active learning. *Journal of Experimental Psychology: Applied* 11(4): 256–265.

Mayer, R. E. and Moreno, R. (2003). Nine ways to reduce cognitive load in multimedia learning. *Educational Psychologist* 38: 45–52.

McCarley, J. S. and Carruth, D. W. (2004). Oculomotor scanning and target recognition in luggage x-ray screening. *International Journal of Cognitive Technology* 9: 26–29.

McCarley, J. S., Kramer, A. F., Wickens, C. D., Vidoni, E. D., and Boot, W. R. (2004). Visual skills in airport security inspection. *Psychological Science* 15: 302–306.

McCarley, J. S., Vais, M. J., Pringle, H., Kramer, A. F., Irwin, D. E., and Strayer, D. L. (2004). Conversation disrupts change detection in complex traffic scenes. *Human Factors* 46: 424–436.

McCarley, J. S., Wang, R. F., Kramer, A. F., Irwin, D. E., and Peterson, M. S. (2003). How much memory does oculomotor search have? *Psychological Science* 14(5): 422–426.

McCarley, J. S., Wiegmann, D., Kramer, A. F., and Wickens, C. D. (2003). Effects of age on utilization and perceived reliability of an automated decision aid in a luggage screening task. In the *Proceedings of the Human Factors and Ergonomics Society 47th Annual Meeting*, Santa Monica, CA.

McCarthy, G. and Donchin, E. (1981). A metric for thought: A comparison of P300 latency and reaction time. *Science* 211: 77–80.

McCoy, C. E. and Mickunas, A. (2000). The role of context and progressive commitment in plan continuation errors. *Proceedings of the IEA 2000/HFES 2000 Congress*, Santa Monica, CA.

McCoy, S. L., Tun, P. A., Cox, L. C., Colangelo, M., Steward, R. A., and Wingfield, A. (2005). Hearing loss and perceptual effort: Down-stream effects on older adults' memory for speech. *Quarterly Journal of Experimental Psychology* 58A: 22–33.

McCracken. J. H. and Aldrich, T. B. (1984). Analyses of selected LHX mission functions: Implications for operator workload and system automation goals (Technical Note ASI479-024-84). Fort Rucker, AL: Army Research Institute Aviation Research and Development Activity.

McDaniel, M. A. and Einstein, G. O. (2007). *Prospective memory: An overview and synthesis of an emerging field*. Thousand Oaks, CA: Sage.

McDougall, S., Tyrer, V., and Folkard, S. (2006). Searching for signs, symbols, and icons: Effects of time of day, visual complexity, and grouping. *Journal of Experimental Psychology: Applied* 12: 118–128.

McDowd, J. M. and Craik, F. I. (1988). Effects of aging and task difficulty on divided attention performance. *Journal of Experimental Psychology: Human Perception and Performance* 14: 267–280.

McFarlane, D. C. (2002). Comparison of four primary methods for coordinating the interruption of people in human-computer interaction. *Human-Computer Interaction* 17: 65–139.

McFarlane, D. C. and Latorella, K. A. (2002). The scope and importance of human interruption in human-computer interface design. *Computer Interaction* 17(1): 1–61.

Meiran, N. (1996). Reconfiguration of processing mode prior to task performance. *Journal of Experimental Psychology: Learning, Memory, and Cognition* 22: 1423–1442.

Metzger, U. and Parasuraman, R (2005). Automation in future air traffic management: effects of decision aid reliability on controller performance and mental workload. *Human Factors* 47: 33–49.

Meyer, D. E. and Kieras, D. E. (1997a). A computational theory of executive cognitive processes and multiple-task performance: Part 1: Basic mechanisms. *Psychological Review* 104: 3–65.

Meyer, D. E. and Kieras, D. E. (1997b). A computational theory of executive cognitive processes and multiple-task performance: Part 2. Accounts of psychological refractory-period phenomena. *Psychological Review* 104: 749–791.

Meyer, J. (2001). Effects of warning validity and proximity on responses to warnings. *Human Factors* 43(4): 563–572.

Meyer, J. (2004). Conceptual issues in the study of dynamic hazard warnings. *Human Factors* 46(2): 196–204.

Meyer, J., Bitan, T., Shinar, D., and Zmora, E. (1999). Scheduling actions and reliance on warnings in a process control task. *Proceedings 1999 Annual Meeting of the Human Factors Society*, Santa Monica, CA.

Miller, G. A. (1956). The magical number seven plus or minus two: Some limits on our capacity for processing information. *Psychological Review* 62: 81–97.

Miller, S. (2002). Effects of rehearsal on interruptions. *Proceedings of the 46th Annual Meeting 2002 Human Factors and Ergonomics Society*.

Molloy, R. and Parasuraman, R. (1996). Monitoring an automated system for a single failure: Vigilance and task complexity effects. *Human Factors* 38: 211–322.

Monk, C., Boehm-Davis, D., and Trafton, J. G. (2004). Recovering from interruptions: Implications for driver distraction research. *Human Factors* 4(46): 650–664.

Monsell, S. (2003). Task switching. *Trends in Cognitive Science* 7: 134–140.

Moray, N. (1959). Attention in dichotic listening: Affective cues and the influence of instructions. *Quarterly Journal of Experimental Psychology* 11: 56–60.

Moray, N (1967). Where is attention limited? A survey and a model. *Acta Psychologica* 27: 84–92.

Moray, N. (1969). *Listening and attention*. Baltimore: Penguin Books.

Moray, N. (1979). *Mental workload: Its theory and measurement*. New York: Plenum Press.

Moray, N. (1986). Monitoring behavior and supervisory control. In Boff, Kaufman, and Thomas, pp. 40.1–40.51.

Moray, N. (1988). Mental workload since since 1979. *International Reviews of Ergonomics* 2: 123–150.

Moray, N. (1997). Process control. In G. S. (Ed.), *Handbook of Human Factors and Ergonomics,* 2d ed. (pp. 1944–1971). Hoboken, NJ: John Wiley.

Moray, N. (2003). Monitoring, complacency, skepticism and eutectic behaviour. *International Journal of Industrial Ergonomics* 31(3): 175–178.

Moray, N., Dessouky, M. I., Kijowski, B. A., and Adapathya, R. (1991). Strategic behavior, workload and performance in task scheduling. *Human Factors* 33, 607–632.

Moray, N. and Rotenberg, I. (1989). Fault management in process control: Eye movements and action. *Ergonomics* 32(11): 1319–1342.

Mori, H. and Hayashi, Y. (1995). Visual interference with users' tasks on multi-window systems. *International Journal of Human-Computer Interaction* 7: 329–340.

Morrow, D. G., Ridolfo, H. E., Menard, W. E., Sanborn, A., Stine-Morrow, E. A. L., Magnor, C., et al. (2003). Environmental support promotes expertise-based mitigation of age differences on pilot communication tasks. *Psychology and Aging* 18: 286–284.

Morrow, D. G., Wickens, C. D., Rantanen, E. M., Chang, D. and Marcus, J. (in press). Designing external aids that support older pilots' communication. *International Journal of Aviation Psychology.*

Mosier, K. L, Skitka, L., Heers, S., and Burdick, M. (1998). Automation bias: Decision making and performance in high-tech cockpits. *International Journal of Aviation Psychology* 8: 47–63.

Most, S. B. and Astur, R. S. (2007). Feature based attentional set as a cause of traffic accidents. *Visual Cognition* 15(2): 125–132.

Mourant, R. R. and Rockwell, T. H. (1972). Strategies of visual search by novice and experienced drivers. *Human Factors* 14: 325–335.

Mulder, G. and Mulder, L. (1981). Information processing and cardiovascular control. *Psychophysiology* 18: 392–401.

Müller, H. J. and Rabbitt, P. M. (1989). Reflexive and voluntary orienting of visual attention: Time course of activation and resistance to interruption. *Journal of Experimental Psychology: Human Perception and Performance* 15: 315–330.

Muthard, E. and Wickens, C. D. (2002). Change detection after preliminary flight decisions: Linking planning errors to biases in plan monitoring. *Proceedings of the 46th Annual Meeting of the Human Factors and Ergonomics Society*, Santa Monica, CA.

Muthard, E. K. and Wickens, C. D. (2005). Display size contamination of attentional and spatial tasks: An evaluation of display minification and axis compression (AHFD-05-12/NASA-05-3). Savoy, IL: University of Illinois, Aviation Human Factors Division

Nagy, A. L. and Sanchez, R. R. (1990). Critical color differences determined with a visual search task. *Journal of the Optical Society of America A* 7: 1209–1217.

Nakayama, K. and Mackeben, M. (1989). Sustained and transient components of focal visual attention. *Vision Research* 29: 1631–1647.

Navalpakkam, V. and Itti, L. (2005). Modeling the influence of task on attention. *Vision Research* 45: 205–231.

Navarro, J. I., Marchena, E., Alcalde, C., Ruiz, G., Llorens, I., and Aguilar, M. Improving attention behaviour in primary and secondary school children with a Computer Assisted Instruction procedure. *International Journal of Psychology.* Vol 38(6), Dec 2003, pp. 359–365

Navon, D. (1984). Resources: A theoretical Soupstone. *Psychological Review* 91: 216–334.

Navon, D. and Gopher, D. (1979). On the economy of the human processing systems. *Psychological Review* 86: 254–255.

Navon, D. and Miller, J. (1987). The role of outcome conflict in dual task interference. *Journal of Experimental Psychology: Human Perception and Performance* 13: 435–448.

Neisser, U. (1963). Decision time without reaction time: Experiments on visual search. *American Journal of Psychology, 76*, 376–395.

Neisser, U. (1967). *Cognitive psychology.* New York: Appleton Century Krofts.

Newman, S. D., Keller, T. A., and Just, M. A. (2007). Volitional control of attention and brain activation in dual task performance. *Human Brain Mapping* 28: 109–117.

Niemela, M. and Saarinen, J. (2000). Visual search for grouped versus ungrouped icons in a computer interface. *Human Factors* 42: 636–659.

Nikolic, M. I., Orr, J. M., and Sarter, N. B. (2004). Why pilots miss the green box: How display context undermines attention capture. *International Journal of Aviation Psychology* 14(1): 39–52.

Nikolic, M. I. and Sarter, N. B. (2001). Peripheral visual feedback. *Human Factors* 43: 30–38.

Nissen, M. J. and Bullemer, P. (1987). Attentional requirements of learning: Evidence from performance measures. *Cognitive Psychology* 19: 1–32.

Nodine, C. F. and Kundel, H. L. (1987). Using eye movements to study visual search and to improve tumor detection. *Radiographics* 7: 1241–1250.

Norman, D. A. and Bobrow, D. G. (1975). On data-limited and resource-limited processes. *Cognitive Psychology* 7: 44–64.

Norman, D. A. and Shallice, T. (1986). Attention to action: Willed and automatic control of behavior. In R. J. Davidson, G. E. Schwartz, and D. Shapiro (Eds.), *Consciousness and self-regulation: Advances in research and theory,* vol. 4 (pp. 1–18). New York: Plenum Press.

North, C. (2006). Information visualization. In Salvendy, pp. 1222–1245.

North, R. and Gopher, D. (1976). Measures of attention as predictors of flight performance. *Human Factors* 18(1): 1–14.

North, R. A. and Riley, V. A. (1989). A predicitive model of operator workload. In McMillan et al., pp. 81–90.

Nothdurft, H. (1992). Feature analysis and the role of similarity in preattentive vision. *Perception and Psychophysics, 52*, 355–375.

Nothdurft, H. C. (2000). Salience from feature contrast: Additivity across dimensions. *Vision Research, 40*, 1183–1201.

Nunes, A., Wickens, C. D., and Yin, S. (2006). Examining the viability of the Neisser search model in the flight domain and the benefits of highlighting in visual search. *Proceedings of the 50th Annual Meeting of the Human Factors and Ergonomics Society*, Santa Monica, CA.

Olmos, O., Liang, C.-C. and Wickens, C.D. (1997). Electronic map evaluation in simulated visual meteorological conditions. *International Journal of Aviation Psychology, 7*(1), 37–66.

Olmos, O., Wickens, C. D., and Chudy, A. (2000). Tactical displays for combat awareness: An examination of dimensionality and frame of reference concepts and the application of cognitive engineering. *International Journal of Aviation Psychology* 10(3): 247–271.

Orasanu, J. and Fischer, U. (1997). Finding decisions in natural environments: The view from the cockpit. In C. Z. G. Klein (Ed.), *Naturalistic decision making* (pp. 343–358). Mahwah, NJ: Lawrence Erlbaum Associates.

O'Regan, J. K., Deubel, H., Clark, J. J., and Rensink, R. A. (2000). Picture changes during blinks: Looking without seeing and seeing without looking. *Visual Cognition* 7: 191–211.

O'Regan, J. K., Rensink, R. A., and Clark, J. J. (1999). Change-blindness as result of "mudsplashes." *Nature* 398: 34.

Owsley, C., Ball, K., McGwin, G., Sloane, M. E., Roenker, D. L., White, M. F., et al. (1998). Visual processing impairment and risk of motor vehicle crash among older adults. *Journal of the American Medical Association* 279: 1083–1088.

Paas, F., Renkl, A., and Sweller, J. (2003). Cognitive load theory and instructional design: Recent developments. *Educational Psychologist* 38: 1–4.

Parasuraman, R. (1986). Vigilance, monitoring and search. In K.R. Boff, L. Kaufman, and J.P. Thomas (Eds.) *Handbook of perception and human performance* Vol. II, New York: Wiley, 43.1–43.39,

Parasuraman, R. (1987). Human computer monitoring. *Human Factors* 29: 695–706.

Parasuraman, R. (2003). Neuroergonomics: Research and practice. *Theoretical Issues in Ergonomics Science* 4: 5–20.

Parasuraman, R. and Davies, R. E. (1984). *Varieties of attention*. New York: Academic Press.

Parasuraman, R. A., Hancock, P., and Olofinboba, O. (1997). Alarm effectiveness in driver-centered collision-warning systems. *Ergonomics* 40(3): 390–399.

Parasuraman, R., Molloy, R., and Singh, I. L. (1993). Performance consequences of automation-induced "complacency." *International Journal of Aviation Psychology* 3(1): 1–23.

Parasuraman, R., Mouloua, M., and Molloy, R. (1996). Effects of adaptive task allocation on monitoring of automated systems. *Human Factors* 38: 665–679.

Parasuraman, R. and Riley, V. (1997). Humans and automation: Use, misuse, disuse, abuse. *Human Factors* 39: 230–253.

Parasuraman, R. and Rizzo, M. (Eds.) (2007). *Neuroergonomics*. Oxford: Oxford University Press.

Pardo, J. V., Pardo, P. J., Janer, K. W., and Raichle, M. E. (1990). The anterior cingulate cortex mediates processing selection in the Stroop attentional conflict paradigm. *Proceedings of the National Academy of Sciences* 87: 256–259.

Park, D. C. (2000). The basic mechanisms accounting for age-related decline in cognitive function. In Park and Schwarz, pp. 3–21.

Park, D. C. and Gutchess, A. H. (2000). Cognitive aging and everyday life. In Park and Schwarz, pp. 217–232.

Park, D. C. and Schwarz, N. (2000). *Cognitive aging: A primer*. Philadelphia: Psychology Press.

Parkes, A. M. and Coleman, N. (1990). Route guidance systems: A comparison of methods of presenting directional information to the driver. In E. J. Lovesey (Ed.), *Contemporary ergonomics 1990* (pp. 480–485). London: Taylor and Francis.

Parks, D. L. and Boucek, G. P., Jr. (1989). Workload prediction, diagnosis, and continuing challenges. In G. McMillan et al. (Eds.). *Applications of human performance models to system design*, New York: Plenum Press, pp. 47–64.

Parks, R. W., Crockett, D. J., Tuokko, H., Beattie, B. L., Ashford, J. W., Coburn, K. L., et al. (1989). Cerebral metabolic effects of a verbal fluency test: A PET scan study. *Journal of Clinical and Experimental Neuropsychology* 10: 565–575.

Pashler, H. (1991). Shifting visual attention and selecting motor responses: Distinct attentional mechanisms. *Journal of Experimental psychology: Human perception and performance, 17*, 1023–1040

Pashler, H., Johnston, J. C., and Ruthruff, E. (2001). Attention and performance. *Annual Review of Psychology* 52: 629–651.

Pashler, H. E. (1989). Dissociations and contingencies between speed and accuracy: Evidence for a two-component theory of divided attention in simple tasks. *Cognitive Psychology* 21: 469–514.

Pashler, H. E. (1998). *The psychology of attention.* Cambridge, MA: MIT Press.

Payne, J. W., Bettman, J. R., and Johnson, E. J. (1993). *The adaptive decision maker.* Cambridge UK: Cambridge University Press.

Perfetti, C. A. and Lesgold, A. M. (1977). Discourse comprehension and sources of individual differences. In M. A. Just and P. A. Carpenter (Eds.), *Cognitive processes in comprehension* (pp. 141–183). Hillsdale, NJ: Erlbaum.

Peters, M. (1981). Attentional asymmetries during concurrent bimanual performance. *Quarterly Journal of Experimental Psychology* 33A: 95–103.

Peterson, M. S., Kramer, A. F., Wang, R. F., Irwin, D. E., and McCarley, J. S. (2001). Visual search has memory. *Psychological Science* 12: 287–292.

Pew, R. W. and Mavor, A. (1998). *Modeling human and organizational behavior.* Washington, DC: National Academy of Sciences.

Plude, D. and Doussard-Rosevelt, J. (1989). Aging, selective attention, and feature integration. *Psychology and Aging* 4: 98–105.

Pohlman, D. L. and Fletcher, J. D. (1999). Aviation personnel selection and training. In D. J. Garland, J. A. Wise, and V. D. Hopkin (Eds.), *Handbook of human factors* (pp. 277–308). Mahwah, NJ: Lawrence Erlbaum.

Polich, J. and Kok, A. (1995). Cognitive and biological determinants of P300: An integrative review. *Biological Psychology* 41: 103–146.

Polson, M. C. and Friedman, A. (1988). Task-sharing within and between hemispheres: A multiple-resources approach. *Human Factors* 30: 633–643.

Pomerantz, J. R. and Pristach, E. A. (1989). Emergent features, attention, and perceptual glue in visual form perception. *Journal of Experimental Psychology* 15: 635–649.

Pope, A. T., Bogart, E. H., and Bartolome, D. (1995). Biocybernetic system evaluates indices of operator engagement. *Biological Psychology* 40: 187–196.

Posner, M. (1978). *Chronometric Explanations of Mind.* Mahwah, NH: Lawrence Erlbaum.

Posner, M. I. (1980). Orienting of attention. *Quarterly Journal of Experimental Psychology* 32: 3–25.

Posner, M. I., Snyder, C. R. R., and Davidson, B. J. (1980). Attention and the detection of signals. *Journal of Experimental Psychology: General* 109(2): 160–174.

Previc, F. H. (1998). The neuropsychology of 3-D space. *Psychological Bulletin* 124: 123–164.

Previc, F. H. (2000, March–April). Neuropsychological guidelines for aircraft control stations. *IEEE Engineering in Medicine and Biology* 81–88.

Previc, F. H. and Ercoline, W. R. E. (2004). *Spatial disorientation in aviation.* Reston, VA: American Institute of Aeronautics and Astronautics, Inc.

Pringle, H. L., Irwin, D. E., Kramer, A. F., and Atchley, P. (2001). The role of attentional breadth in perceptual change detection. *Psychonomic Bulletin and Review* 8: 89–95.

Prinzel, L. J., Freeman, F. G., Scerbo, M. W., Mikulka, P. J., and Pope, A. T. (2000). A closed-loop system for examining psychophysiological measures for adaptive task allocation. *International Journal of Aviation Psychology* 10: 393–410.

Pritchard, W. S. (1981). Psychophysiology of P300. *Psychological Bulletin, 89,* 506–540.

Pritchett, A. (2001). Reviewing the role of cockpit alerting systems. *Human Factors of Aerospace Safety* 1: 5–38.

Proctor, R. and Proctor, J. (2006). Sensation and perception. In Salvendy, pp. 53–57.

Quinlan, P. T. (2003). Visual feature integration theory: Past, present, and future. *Psychological Bulletin* 129: 643–673.

R. lee, M. J., R., and Caretta, T. R. (1996). Central role of g in military pilot selection. *International Journal of Aviation Psychology* 6: 111–123.

Rabbitt, P. M. (1991). Mild hearing loss can cause apparent memory failures which increase with age and reduce with IQ. *Acta Otolaryngolocica, Suplementum* 476: 167–176.

Rabbitt, P. M. A. (1968). Channel capacity, intelligibility and immediate memory. *Quarterly Journal of Experimental Psychology* 20: 241–248.

Raby, M. and Wickens, C. D. (1994). Strategic workload management and decision biases in aviation. *International Journal of Aviation Psychology* 4(3): 211–240.

Ranney, T. A., Harbluk, J. L., and Noy, Y. I. (2005). Effects of voice technology on test track driving performance: Implications for driver distraction. *Human Factors* 47: 439–454.

Rantanen, E. M., Wickens, C. D., Xu, X., and Thomas, L. C. (2003). Developing and validating human factors certification criteria for cockpit displays of traffic information avionics. Aviation Human Factors Division TR AFD04-01, FAA 04-01, University of Illinois, Technical Report.

Rauschenberger, R. and Yantis, S. (2006). Perceptual encoding efficiency in visual search. *Journal of Experimental Psychology: General* 135: 116–131.

Rayner, K. (1998). Eye movements in reading and information processing: 20 years of research. *Psychological Bulletin* 124(3): 372–422.

Reason, J. T. (1990). *Human error*. Cambridge, UK: Cambridge University Press.

Recarte, M. A. and Nunes, L. M. (2000). Effects of verbal and spatial-imagery tasks on eye fixations while driving. *Journal of Experimental Psychology: Applied* 6: 31–43.

Ree, M.J. and Caretta, T.R. (1996). The central role of g in military pilot selection. *International Journal of Aviation Psychology, 6,* 111–123.

Regan, D. (2000). *Human perception of objects*. Sunderland, MA: Sinauer Associates.

Reichle, E. D., Carpenter, P. A., and Just, M. A. (2000). The neural basis of strategy and skill in sentence-picture verification. *Cognitive Psychology* 40: 261–295.

Reid, G. B. and Nygren, T. E. (1988). The subjective workload assessment technique: A scaling procedure for measuring mental workload. In P. A Hancock and N. Meshkati (Eds.), *Human mental workload* (pp. 185–213). Amsterdam, NL: North Holland.

Reising, J., Liggett, K., Kustra, T. W., Snow, M. P., Hartsock, D. C., and Barry, T. P. (1998). Evaluation of pathway symbology used to land from curved approaches. *Proceedings of the 42nd Annual Meeting of the Human Factors and ErgonomicsSociety*, Santa Monica, CA.

Remington, R. W., Johnston, J. C., Ruthruff, E., Gold, M., and Romera, M. (2001). Visual search in complex displays: Factors affecting conflict detection by air traffic controllers. *Human Factors* 42: 349–366.

Rensink, R. A. (2002). Change detection. *Annual Review of Psychology* 53: 245–277.

Rensink, R. A., O'Regan, J. K., and Clark, J. J. (1997). To see or not to see: The need for attention to perceive changes in scenes. *Psychological Science* 8: 368–373.

Reynolds, D. (1966). Time and event uncertainty in unisensory reaction time. *Journal of Experimental Psychology* 71: 286–293.

Reynolds, J. H. and Desimone, R. (2003). Interacting roles of attention and visual salience in V4. *Neuron* 37: 853–863.

Richland, L. E., Linn, M. C., and Bjork, R. A. (2007). Instruction. In Durso et al., pp. 555–584.

Roberts, R. J., Hager, L. D., and Heron, C. (1994). Prefrontral cognitive processes: Working memory and inhibition in the antisaccade task. *Journal of Experimental Psychology: General* 123: 347–393.

Roenker, D. L., Cissell, G. M., Ball, K., Wadley, V. G., and Edwards, J. D. (2003). Speed-of-processing and driving simulator training result in improved driving performance. *Human Factors* 45: 218–233.

Rogers, R. D. and Monsell, S. (1995). Costs of a predictable switch between simple cognitive tasks. *Journal of Experimental Psychology: General* 124(2): 207–231.

Rollins, R. A. and Hendricks, R. (1980). Processing of words presented simultaneously to eye and ear. *Journal of Experimental Psychology: Human Perception and Performance* 6: 99–109.

Rosen, V. M. and Engle, R. W. (1997). The role of working memory capacity in retrieval. *Journal of Experimental Psychology: General* 126: 211–227.

Rouse, W. V. (1988). Adaptive aiding for human computer control. *Human Factors* 30(4): 431–443.

Rubinstein, J. S., Meyer, D. E., and Evans, J. E. (2001). Executive control of cognitive processes in task switching. *Journal of Experimental Psychology: Human Perception and Performance* 27(4): 763–797.

Ruggerio, F. and Fadden, D.M. (1987). Pilot subjective evaluation of workload during a flight test certification program. In A. H. Roscoe (Ed.), *The Practical Assessment of Pilot Workload*. NATO AGARDograph no. AGARD-AG-282. Essex, UK: Specialised Printing Services LTD.

Russo, J. E. (1977). The value of unit price information. *Journal of Marketing Research* 14: 193–201.

Salthouse, T. A. (1996). The processing-speed theory of adult age differences in cognition. *Psychological Review* 103(3): 403–428.

Salvendy, G. (Ed.) (2006). *Handbook of Human Factors and Ergonomics*, 3d ed. Hoboken, NJ: John Wiley and Sons.

Salvucci, D., and Taatgen (2008, in press). Threaded Cognition. *Psychological Review, 131*.

Sanders, A. F. (1963). *The Selective Process in the Functional Visual Field*. Assen, Netherlands: Van Gorcum.

Sanders, A. F. and Houtmans, M. (1985). Perceptiula processing modes in the functional visual field. *Acta Psychologica* 58: 251–261.

Sanderson, P. M., Flach, J. M., Buttigieg, M. A., and Casey, E. J. (1989). Object displays do not always support better integrated task performance. *Human Factors* 31: 183–198.

Sarno, K. J. and Wickens, C. D. (1995). The role of multiple resources in predicting time-sharing efficiency: An evaluation of three workload models in a multiple task setting. *International Journal of Aviation Psychology* 5(1): 107–130.

Sarter, N., Mumaw, R., and Wickens, C. D. (2007). Pilots' monitoring strategies and performance on highly automated flight decks: an empirical study combining behavioral and eye tracking data. *Human Factors, 49*, 347–356.

Sarter, N. B. (2007). Multiple resource theory as a basis of multimodal interface design. In Kramer, Wiegmann, and Kirlik, pp. 187–195.

Sarter, N. B. and Woods, D. D. (1997). Team play with a powerful and independent agent: Operational experiences and automation surprises on the Airbus A-320. *Human Factors* 39(4): 553–569.

Scerbo, M. W. (2001). Adaptive automation. In W. Karwowski (Ed.), *International Encyclopedia of Ergonomics and Human Factors* (pp. 1077–1079). London: Taylor and Francis.

Scerbo, M. W. (2007). Adaptive automation. In Parasuraman and Rizzo, pp. 239–252.

Scerbo, M. W., Freeman, F. G., and Mikulka, P. J. (2003). A brain-based system for adaptive automation. *Theoretical Issues in Ergonomics Science* 4: 200–219.

Schmidt, R. (1988*). Motor Control and Learning: A Behavioral Emphasis*. Champaign, IL: Human Kinetics.

Schmidt, R. and Bjork, R. A. (1992). New conceptualizations of practice: Common principles in three paradigms suggest new concepts for training. *Psychological Science* 3: 207–217.

Schneider, W. (1985). Training high-performance skills: Fallacies and guidelines. *Human Factors* 27(3): 285–300.

Schneider, W. and Detweiller, M. (1988). The role of practice in dual task performance. *Human Factors* 30(5): 539–556.

Schneider, W. and Fisk, A. F. (1982). Concurrent automatic and controlled visual search *Journal of Experimental Psychology: Learning, Memory and Cognition* 8: 261–278.

Schneider, W. and Shiffrin, R. (1977). Controlled and automatic human information processing I: Detection, search and attention. *Psychological Review* 84: 1–66.

Schons, V. and Wickens, C. (1993). Visual separation and information access in aircraft display layout. Aviation Research Lab Technical Report ARL 93-7/NASA A31-93-1, University of Illinois.

Schrmorrow, D., Stanney, K. M., Wilson, G., and Young, P. (2006). Augmented cognition in human-system interaction. In G. Salvendy (Ed.), *Handbook of Human Factors and Ergonomics*. Hoboken NJ: John Wiley and Sons.

Schroeder, D. H. and Salthouse, T. A. (2004). Age-related effects on cognition between 20 and 50 years of age. *Personality and Individual Differences* 36: 393–404.

Schumacher, E.H, Seymour, T.L., Glass, J.M., Fencsik, D.E., Lauber, E.J., Kieras, D.E., and Meyer, D.E. (2001). Virtually perfect time sharing in dual task performance: Uncorking the central cognitive bottleneck. *Psychological Science. 12*, 1–108.

Schutte, P. C. and Trujillo, A. C. (1996). Flight crew task management in non-normal situations. *Proceedings of the 40th Annual Meeting of the Human Factors and Ergonomics Society*, Santa Monica, CA.

Schvaneveldt, R.W., (1969). Effects of complexity in simultaneous reaction time tasks. *Journal of Experimental Psychology, 81*, 289–296.

Scialfa, C. T., Jenkins, L., Hamaluk, E., and Skaloud, P. (2000). Aging and the development of automaticity in conjunction search. *Journal of Gerontology: Psychological Sciences* 55B: 27–46.

Scialfa, C. T., Kline, D. W., and Lyman, B. J. (1987). Age differences in target identification as a function of retinal location and noise level: Examination of the useful field of view. *Psychology and Aging* 2: 14–19.

Seagull, F. J. and Gopher, D. (1997). Training head movement in visual scanning: An embedded approach to the development of piloting skills with helmet-mounted displays. *Journal of Experimental Psychology: Applied* 3(3): 163–180.

Seagull, F. J., et al. (2000). Auditory Alarms: from alerting to informing. Proceedings of the 2000 meeting of the International Ergonomics Society and the Human Factors Society, Santa Monica, CA.

Seagull, F.J., Xiao, Y, and Plasters, C. (2004). Information accuracy and sampling effort: A field study of surgical scheduling coordination. *IEEE Transactions in Systems: Man and Cybernetics. Part A: Systems and Humans* 34 6; 764–771.

Seidler, K. and Wickens, C. D. (1992). Distance and organization in multifunction displays. *Human Factors* 34: 555–569.

Sekuler, R. and Ball, K. (1986). Visual localization: Age and practice. *Journal of the Optical Society of America* A3: 864–867.

Sem-Jacobson, C. W. (1959). Electroencephalographic study of pilot stress. *Aerospace Medicine* 30: 797–803.

Sem-Jacobson, C. W. and Sem-Jacobson, J. E. (1963). Selection and evaluation of pilots for high performance aircraft and spacecraft by in-flight EEG study of stress tolerance. *Aerospace Medicine* 34: 605–609.

Senders, J. (1964). The human operator as a monitor and controller of multidegree of freedom systems. *IEEE Transactions on Human Factors in Electronics* HFE-5: 2–6.

Senders, J. (1980). *Visual scanning processes.* Unpublished doctoral dissertation, University of Tilburg, The Netherlands.

Shallice, T. and Burgess, P. (1993). Supervisory control of action and thought selection. In A. Baddeley and L. Weiskrantz (Eds.), *Attention: Selection, awareness, and control* (pp. 171–187). Oxford: Oxford University Press.

Shallice, T., McLeod, P., and Lewis, K. (1985). Isolating cognitive modules with the dual-task paradigm: Are speech perception and production modules separate? *Quarterly Journal of Experimental Psychology* 37: 507–532.

Shapiro, K. L. and Raymond, J. (1989). Training of efficient oculomotor strategies enhances skill acquisition. *Acta Psychologica* 71: 217–242.

Sharit, J. (2006). Human error. In Salvendy, pp. 708–760.

Sheridan, T. (1970). On how often the supervisor should sample. *IEEE Transactions on Systems Science and Cybernetics, SSC-6*(2): 140–145.

Sheridan, T. (1997). Supervisory control. In G. Salvendy (Ed.), *Handbook of human factors and ergonomics*, 2d ed. (pp. 1295–1327). City: Wiley Interscience.

Sheridan, T. (2002). *Humans and Automation.* Chichester, UK: John Wiley and Sons.

Sheridan, T. and Parasuraman, R. (2006). Human–automation interaction. In R. Nickerson (Ed.), *Reviews of Human Factors and Ergonomics* (pp. 89–129). Santa Monica, CA: Human Factors.

Shiu, L. and Pashler, H. (1994). Negligible effect of spatial precuing on identification of single digits. *Journal of Experimental Psychology: Human Perception and Performance* 20: 1037–1054.

Shugan, S. M. (1980). The cost of thinking. *Journal of Consumer Research* 7: 99–111.

Shute, V. J. (1991). Who is likely to acquire programming skills? *Journal of Educational Computing Research* 7: 1–24.

Simons, D. J. and Chabris, C. F. (1999). Gorillas in our midst: Sustained inattentional blindness for dynamic events. *Perception* 28: 1059–1074.

Simons, D. J. and Levin, D. T. (1997). Change blindness. *Trends in Cognitive Science* 1(7): 261–267.

Simons, D. J. and Levin, D. T. (1998). Failure to detect changes to people during a real-world interaction. *Psychonomic Bulletin and Review* 5(4): 644–649.

Sirevaag, E., Kramer, A., Wickens, C., Reisweber, M., Strayer, D., and Grenell, J. (1993). Assessment of pilot performance and mental workload in rotary wing aircraft. *Ergonomics* 9: 1121–1140.

Sit, R. A. and Fisk, A. D. (1999). Age-related performance in a multiple-task environment. *Human Factors* 41: 26–34.

Sklar, A. F. and Sarter, N. B. (1999). Good vibrations: The use of tactile feedback to support mode awareness on advanced technology aircraft. *Human Factors* 41(4): 543–552.

Slamecka, N. J. and Graf, P. (1978). The generation effect: Delineation of a phenomena. *Journal of Experimental Psychology: Human Learning and Memory* 4: 592–604.

Smith, J. D., Redford, J. S., Gent, L., and Washburn, D. A. (2005). Visual search and the collapse of categorization. *Journal of Experimental Psychology: General* 134: 443–460.

Smith, M. (1967). Theories of the psychological refractory period. *Psychological Bulletin* 19: 352–359.

Smith, P., Bennett, K., and Stone, B. (2007). Representational aiding to support performance on problem solving tasks. In R. Williges (Ed.), *Review of Human Factors and Ergonomics*, vol. 2 (pp. 74–108). Santa Monica, CA: Human Factors.

Sorkin, R. D. (1989). Why are people turning off our alarms? *Human Factors Bulletin* 32(4): 3–4.

Sorkin, R. D., Kantowitz, B. H., and Kantowitz, S. C. (1988). Likelihood alarm displays. *Human Factors* 30(4): 445–459.

Sorkin, R. D. and Woods, D. D. (1985). Systems with human monitors, a signal detection analysis. *Human-Computer Interaction* 1, 49–75.

Spector, A. and Biederman, I. (1976). Mental set and mental shift revisited. *Journal of Psychology* 89: 669–679.

Spelke, E., Hirst, W., and Neisser, U. (1976). Skills of divided attention. *Cognition* 4: 215–230.

Spence, C. J. and Driver, J. (1994). Covert spatial orienting in audition: Exogenous and endogenous mechanisms. *Journal of Experimental Psychology: Human Perception and Performance, 20*, 555–574.

Spence, C. and Driver, J. (1996). Audiovisual links in endogenous covert spatial attention. *Journal of Experimental Psychology: Human Perception and Performance* 22: 1005–1030.

Spence, C. and Driver, J. (1997). Cross-modal links in attention between audition, vision, and touch: Implications for interface design. *International Journal of Cognitive Ergonomics* 1: 351–373.

Spence, C., McDonald, J., and Driver, J. (2004). Exogenous spatial-cuing studies of human crossmodal attention and multisensory integration. In C. Spence and J. Driver (Eds.), *Crossmodal space and crossmodal attention* (pp. 277–320). Oxford: Oxford University Press.

Spence, C., Pavani, F., and Driver, J. (2000). Crossmodal links between vision and touch in covert endogenous spatial attention. *Journal of Experimental Psychology: Human Perception and Performance* 26: 1298–1319.

Spence, C., Ranson, J., and Driver, J. (2000). Crossmodal selective attention: On the difficulty of ignoring sounds at the locus of visual attention. *Perception and Psychophysics* 62: 410–424.

Spence, C. and Read, L. (2003). Speech shadowing while driving: On the difficulty of splitting attention between eye and ear. *Psychological Science* 14: 251–256.

Spence, C. and Walton, M. (2005). On the inability to ignore touch when responding to vision in the crossmodal congruency task. *Acta Psychologica* 118: 47–70.

Spence, C. J. and Driver, J. (1994). Covert spatial orienting in audition: Exogenous and endogenous mechanisms. *Journal of Experimental Psychology: Human Perception and Performance* 20: 555–574.

Sperling, G. and Melchner, M. (1978). Visual search, visual attention, and the attention operating characteristic. In J. Requin (Ed.), *Attention and performance VIII* (pp. 675–686). Hillsdale, NJ: Erlbaum.

Srinivasan, R. and Jovanis, P. R. (1997). Effect of in-vehicle route guidance systems on driver workload and choice of vehicle speed: Findings from a driving simulator experiment. In Y. I. Noy (Ed.), *Ergonomics and safety of intelligent driver interfaces* (pp. 97–114). Mahwah, NJ: Lawrence Erlbaum.

St. John, M. and Manes, D. I. (2002). Making unreliable automation useful. *Proceedings of the 46th Annual Human Factors and Ergonomic Society*, Santa Monica (CDRom).

Stagar, P. and Angus, R. (1978). Locating crash sites in simulated air-to-ground visual search. *Human Factors, 20,* 453–466.

Stankov, L. (1983). Attention and intelligence. *Journal of Educational Psychology* 74(4): 471–490.

Stankov, L. (1988). Single tasks, competing tasks, and their relationship to the broad factors of intelligence. *Personality and Individual Difference* 9: 25–44.

Stanton, N. (1994). *Human factors in alarm design.* London: Taylor and Francis.

Stanton, N.A. and Young, M.S. (1998). Vehicle automation and driving performance. *Ergonomics, 41,* 1014–1028.

Steblay, N. M. (1992). A meta-analytic review of the weapon focus effect. *Law and Human Behavior* 16: 413–434.

Stelter E. M. and Wickens, C. D. (2006). Pilots strategically compensate for display enlargements in surveillance and flight control tasks. *Human Factors* 48(1): 166–181.

Sterman, B. and Mann, C. (1995). Concepts and applications of EEG analysis in aviation performance evaluation. *Biological Psychology* 40: 115–130.

Sterman, B., Mann, C., Kaiser, D., and Suyenobu, B. (1994). Multiband topographic EEG analysis of a simulated visuo-motor aviation task. *International Journal of Psychophysiology* 16: 49–56.

Stern, J. A., Boyer, D., and Schroeder, D. (1994). Blink rate: A possible measure of fatigue. *Human Factors* 36: 285–297.

Sternberg, S. (1966). High-speed scanning in human memory. *Science, 153,* 652–654.

Sternberg, S., ((1969). The discovering of processing stages: Extension of Donder's method. *Acta Psychologica. 30,* 276–315.

Stokes, A. F., Wickens, C. D., and Kite, K. (1990). *Display technology: Human factors concepts.* Warrendale, PA: Society of Automotive Engineers.

Stone, E.R., Yates, J.F., and Parker, A., (1997) Effects of numerical and graphical displays on professed risk-taking behavior. *Journal of Experimental Psychology: Applied. 3,* 243–256.

Strayer, D. L. and Drews, F. A. (2007). Multitasking in the automobile. In Kramer, Wiegmann, and Kirlik, pp. 121–133.

Strayer, D. L., Drews, F. A., and Johnston, W. A. (2003). Cell phone-induced failures of visual attention during simulated driving. *Journal of Experimental Psychology, Applied* 9: 23–32.

Strayer, D. L. and Johnston, W. A. (2001). Driven to distraction: Dual-task studies of driving and conversing on a cellular telephone. *Psychological Science* 12(6): 462–466.

Stroop, J. R. (1992). Studies of interference in serial verbal reactions. *Journal of Experimental Psychology: General* 121: 15–23.

Summala, H., Pasanen, E., Räsänen, M., and Sievänen, J. (1996). Bicycle accidents and drivers' visual search at left hand right turns. *Accident Analysis and Prevention* 28: 147–153.

Summers, J. J. (1989). Motor programs. In D. H. Holding (Ed.), *Human skills,* 2d ed. New York: Wiley.

Sutton, S., Braren, M., Zubin, J., and John, E. R. (1965). Evoked-potential correlates of stimulus uncertainty. *Science* 150: 1187–1188.

Sverko, B. (1977). Individual differences in time sharing performance. *Acta Institute Psychologica* 79: 17–30.

Sweller, J. (1994). Cognitive load theory: Learning difficulty and instructional design. *Learning and Instruction* 12(3): 295–312.

Sweller, J. and Cooper, G.A. (1985). The use of worked examples as a substitute for problem solving in learning algebra. *Cognition and Instruction* 2: 59–89.

Sweller, J. and Chandler, P. (1994). Why some material is difficult to learn. *Cognition and Learning* 12(3): 185–233.

Swets, J. A. and Pickett, R. M. (1982). *The evaluation of diagnostic systems*. New York: Academic Press.

Teichner, W. H. and Mocharnuk, J. B. (1979). Visual search for complex targets. *Human Factors* 21: 259–275.

Telford, C. W. (1931). Refractory phase of voluntary and associate response. *Journal of Experimental Psychology* 14: 1–35.

Tenney, Y. and Pew, R. W. (2007). Situation awareness. In R. Williges (Ed.), *Reviews of human factors and ergonomics*, vol. 2 (pp. 1–34). Santa Monica, CA: Human Factors.

Theeuwes, J. (1994). Endogenous and exogenous control of visual attention. *Perception* 23: 429–440.

Theeuwes, J. (1996). Visual search at intersections: An eye-movement analysis. In A. G. Gale (Ed.), *Vision in vehicles 5* (pp. 203–212). Amsterdam: Elsevier.

Theeuwes, J. (2000). Commentary on Räsänen and Summala, "Car drivers' adjustments to cyclists at roundabouts." *Transportation Human Factors* 2: 19–22.

Theeuwes, J. and Hagenzieker, M. P. (1993). Visual search of traffic scenes: On the effect of location expectation. In A. G. Gale et al. (Eds.), *Vision in vehicles IV*. Amsterdam. Elsevier. 149–159.

Theeuwes, J., Kramer, A. F., Hahn, S., and Irwin, D. E. (1998). Our eyes do not always go where we want them to go: Capture of the eyes by new objects. *Psychological Science* 9: 379–385.

Thomas, L. C. and Rantanen, E. M. (2006). Human Factors issues in implementing of advanced aviation technologies: A case of false alerts and cockpit displays of traffic information. *Theoretical Issues in Ergonomic Science* 7(5): 501–523.

Thomas, L. C. and Wickens, C. D. (2006). Effects of battlefield display frames of reference on navigational tasks, spatial judgments and change detection. *Ergonomics* 49, 1154–1173.

Tindall-Ford, S., Chandler, P., and Sweller, J. (1997). When two sensory modes are better than one. *Journal of Experimental Psychology, Applied* 3(4): 257–287.

Titchner, E. B. (1908). *Lectures on the elementary psychology of feeling and attention*. New York: MacMillan Publishing Inc.

Todd, P. M. and Gigerenzer, G. (2000). Précis of simple heuristics that make us smart. *Behavioral and Brain Sciences* 23: 727–780.

Townsend, J. T. (1972). Some results concerning the identifiability of parallel and serial processes. *British Journal of Mathematical and Statistical Psychology, 25,* 168–199.

Townsend, J. T. and Ashby, F. G. (1983). *Stochastic modeling of elementary psychological processes*. Cambridge, UK: Cambridge University Press.

Townsend, J. T. and Wenger, M. J. (2004). The serial-parallel dilemma: A case study in a linkage of theory and method. *Psychonomic Bulletin and Review* 11: 391–418.

Trafton, J. G. (2007). Dealing with Interruptions. *Reviews of Human Factors and Ergonomics 3*. Santa Monica, CA: Human Factors.

Trafton, J. G., Altman, E., and Brock, D. (2005). Huh? What was I doing? *Proceedings of the 49th Conference of Human Factors and Ergonomics Society*, Santa Monica, CA.

Trafton, J. G., Altmann, E. M., Brock, D. P. and Mintz, F. E. (2003). Preparing to resume an interrupted task: Effects of prospective goal encoding and retrospective rehearsal. *International Journal of Human Computer Studies.* 58(5), 582–602.

Treisman, A. (1964a). The effect of irrelevant material on the efficiency of selective listening. *American Journal of Psychology* 77: 533–546.

Treisman, A. (1964b).Verbal cues, language and meaning in attention. *American Journal of Psychology* 77: 206–214.

Treisman, A. (1982). Perceptual grouping and attention in visual search for features and objects. *Journal of Experimental Psychology: Human Perception and Performance* 8: 194–214.

Treisman, A. (1986). Properties, parts, and objects. In Boff, Kaufman, and Thomas, pp. 31–31, 35–70.

Treisman, A. and Davies A. (1973). Divided attention to eye and ear. In S. Kornblum (Ed.), *Attention and performance IV* (pp. 101–118). New York: Academic Press.

Treisman, A. M. and Gelade, G. (1980). A feature-integration theory of attention. *Cognitive Psychology* 12: 97–136.

Treisman, A. and Gormican, S. (1988). Feature analysis in early vision: Evidence from search asymmetries. *Psychological Review* 95: 15–48.

Treisman, A. and Sato, S. (1990). Conjunction search revisited. *Journal of Experimental Psychology: Human Perception and Performance* 16: 459–478.

Treisman, A. and Schmidt, H. (1982). Illusory conjunctions in the perception of objects. *Cognitive Psychology* 14: 107–141.

Treisman, A. and Souther, J. (1985). Search asymmetry: A diagnostic for preattentive processing of separable features. *Journal of Experimental Psychology: General* 114: 285–310.

Tripp, L. D. and Warm, J. S. (2007). Transcranial Doppler sonography. In Parasuraman and Rizzo, pp. 82–94.

Tsang, P. (2006). Regarding time-sharing with convergent operations. *Acta Psychologica* 121: 137–175.

Tsang, P. and Vidulich, M. A. (2006). Mental workload and situation awareness. In Salvendy, pp. 243–268.

Tsang, P. S. and Shaner, T. L. (1998). Age, attention, expertise, and time-sharing performance. *Psychology and Aging* 13: 323–347.

Tsang, P. S. and Wickens, C. D. (1988). The structural constraints and strategic control of resource allocation. *Human Performance* 1: 45–72.

Tsimhoni, O., Smith, D., and Green, P. (2004). Address entry while driving: speech recognition versus touch screen keyboard. *Human Factors* 46(4): 600–610.

Tufte, E. (2001). *The visual display of quantitative information*, 2d ed. Cheshire, CT: Graphics Press.

Tulga, M. K. and Sheridan, T. B. (1980). Dynamic decisions and workload in multitask supervisory control. *IEEE Transactions on Systems, Man and Cybernetics* SMC-10: 217–232.

Tun, P. A., O'Kane, G., and Wingfield, A. (2002). Distraction by competing speech in young and older adult listeners. *Psychology and Aging* 17: 453–467.

Tun, P. A. and Wingfield, A. (1999). One voice too many: Adult age differences in language processing with different types of distracting sounds. *Journal of Gerontology: Psychological Sciences* 54B: P317–P327.

Turner, M. L. and Engle, R. W. (1989). Is working memory capacity task dependent? *Journal of Memory and Language* 28: 127–154.

Tversky, A. (1972). Elimination by aspects: A theory of choice. *Psychological Review* 79: 281–299.

Tversky, A. and Kahneman, D. (1974). Judgment under uncertainty: Heuristics and biases. *Science* 185: 1124–1131.

Umiltà, C. and Moscovitch, M. (Eds.). *Attention and performance XV: Conscious and nonconscious information processing*. Cambridge, MA: MIT Press.

Van Erp, J. (2007). Tactile displays for navigation and orientation. Dissertation. Utrecht University, Leiden, NL: Mostert and Van Onderen.

Van Erp, J., Veltman, J. A., and van Vern, H. A. (2003). A tactile cockpit instrument to support altitude control. *Human Factors and Ergonomics Society 47th Annual Meeting*, Santa Monica, CA.

Venturino, M. (1991). Automatic processing, code dissimilarity and the efficiency of successive memory searches. *Journal of Experimental Psychology: Human Perception and Performance* 17: 677–695.

Verhaeghen, P. (2000). The parallels in beauty's brow: Time-accuracy functions and their implications for cognitive aging theories. In T. J. Perfect and E. A. Maylor (Eds.), *Models of cognitive aging* (pp. 51–86). New York: Oxford University Press.

Verhaeghen, P. and Cerella, J. (2002). Aging, executive control, and attention: A review of meta-analyses. *Neuroscience and Biobehavioral Reviews* 26: 849–857.

Verhaeghen, P., Steitz, D. W., Sliwinski, M. J., and Cerella, J. (2003). Aging and dual-task performance: A meta-analysis. *Psychology and Aging* 18: 443–460.

Verwey, W. B. and Veltman, H. A. (1996). Detecting short periods of elevated workload: A comparison of nine workload assessment techniques. *Journal of Experimental Psychology: Applied* 2(3): 270–285.

Vicente, K. and Rasmussen, J. (1992). Ecological interface design: Theoretical foundations. *IEEE Transactions on Systems, Man, and Cybernetics* 22: 589–606.

Vicente, K. J. (2002). Ecological interface design: Progress and challenges. *Human Factors* 44: 62–78.

Vicente, K. J., Thornton, D. C., and Moray, N. (1987). Specral analysis of sinus arrythmia: A measure of mental effort. *Human Factors* 29(2): 171–182.

Vidulich, M. A. (1988). Speech responses and dual task performance: Better time-sharing or asymmetric transfer. *Human Factors* 30: 517–534.

Vidulich, M. A. and Wickens, C. D. (1985). Stimulus-central processing-response compatibility: Guidelines for the optimal use of speech technology. *Behavior Research Methods, Instruments, and Computers* 17(2): 243–249.

Vidulich, M. A. and Wickens, C. D. (1986). Causes of dissociation between subjective workload measures and performance. *Applied Ergonomics* 17: 291–296.

Vincow, M. A. and Wickens, C. D. (1993). Spatial layout of displayed information: Three steps toward developing quantitative models. *Proceedings of the 37th Annual Meeting of the Human Factors Society*, Santa Monica, CA.

Wachtel, P. L. (1967). Conceptions of broad and narrow attention. *Psychological Bulletin* 68: 417–419.

Wainer, H. and Thissen, D. (1981). Graphical data analysis. *Annual Review of Psychology* 32: 191–241.

Walker, N. and Fisk, A. D. (1995). Human factors goes to the grid iron: Developing a quarterback training system. *Ergonomics in Design, 3(2)*, 8–13.

Wallis, G. and Bülthoff, H. (2000). What's scene and not seen: Influences of movement and task upon what we see. *Visual Cognition* 7: 175–190.

Wallsten, T.S. and Barton, C., (1982). Processing probabilitistic multidimensional information for decisions. *Journal of Experimental Psychology: Learning Memory and Cognition, 8*, 361–384.

Wang, M. J. J., Lin, S. C., and Drury, C. G. (1997). Training for strategy in visual search. *International Journal of Industrial Ergonomics* 20: 101–108.

Wang, Q. and Cavanagh, P. (1994). Familiarity and pop-out in visual search. *Perception and Psychophysics* 56: 495–500.

Warm, J. (1984). *Sustained attention in human performance*. Chichester, UK: John Wiley and Sons.

Warm, J., Denber, W., and Hancock, P. A. (1996). Vigilance and workload in automated systems. In R. Parasuraman and M. Moloua (Eds.), *Automation and Human Performance: Theory and Applications.* Mahwah, NJ: Erlbaum.

Watson, M. and Sanderson, P. (2004). Sonification supports eyes-free respiratory monitoring and task time-sharing. *Human Factors* 46: 497–517.

Weinstein, L. F. and Wickens, C. D. (1992). Use of nontraditional flight displays for the reduction of central visual overload in the cockpit. *International Journal of Aviation Psychology* 2: 121–142.

Weise, E. E. and Lee, J. D (2007). Attention grounding: A new approach to in-vehicle information system implementation. *Theoretical Issues in Ergonomic Sciences* 8: 255–276.

Welford, A. T. (1952). The psychological refractory period and the timing of high speed performance. *British Journal of Psychology* 43: 2–19.

Welford, A. T. (1967). Single channel operation in the brain. *Acta Psychologica* 27: 5–21.

Welford, A. T., (1968). *Fundamentals of Skill.* London U.K.: Metheun Press.

Weltman, G., Smith, J. E., and Edstrom, G. H. (1971). Perceptual narrowing during simulated pressure-chamber exposure. *Human Factors* 13: 99–107.

Wetherell, A. (1979). Short term memory for verbal and graphic route information. In *Proceedings 29th Annual Meeting of the Human Factors Society.* Santa Monica CA: Human Factors.

Wickens, C. D. (1976). The effects of divided attention on information processing in tracking. *Journal of Experimental Psychology: Human Perception and Performance* 2: 1–13.

Wickens, C. D. (1980). The structure of attentional resources. In R. Nickerson (Ed.), *Attention and performance VIII* (pp. 239–258). Hillsdale, NJ: Lawrence Erlbaum.

Wickens, C. D. (1984). Processing resources in attention. In Parasuraman and Davies, pp. 63–101.

Wickens, C. D. (1986). The effects of control dynamics on performance. In Boff, Kaufman, and Thomas, pp. 39.1–39.43.

Wickens, C. D. (1991). Processing resources and attention. In Damos, pp. 3–34.

Wickens, C. D. (1992). Virtual reality and education. In *Proceedings of the IEEE International Conference on Systems, Man, and Cybernetics,* vol. 1 (pp. 842–847). New York: IEEE.

Wickens, C. D. (1993). Cognitive factors in display design. *Journal of the Washington Academy of Sciences* 83(4): 179–201.

Wickens, C. D. (1998). Common sense statistics. *Ergonomics in Design* October: 18–22.

Wickens, C. D. (1999a). Aerospace Psychology. In P. Hancock (Ed.), *Human performance and ergonomics: Perceptual and cognitive principles.* San Diego, CA: Academic Press.

Wickens, C. D. (1999b). Frames of reference for navigation. In Gopher and Koriat, pp. 113–144.

Wickens, C. D. (2000). Human factors in vector map design: The importance of task-display dependence. *Journal of Navigation* 53(1): 54–67.

Wickens, C. D. (2001). Attention to safety and the psychology of surprise. In R. Jensen and L Rakovan (Eds.), *Proceedings of the 1st International Symposium on Aviation Psychology* (pp. 1–11). Columbus: Ohio State University Press.

Wickens, C. D. (2002). Multiple resources and performance prediction. *Theoretical Issues in Ergonomics Science* 3(2): 159–177.

Wickens, C. D. (2003). Pilot actions and tasks: Selections, execution, and control. In P. Tsang and M. Vidulich (Eds.), *Principles and Practices of Aviation Psychology* (pp. 239–263). Mahwah, NJ: Lawrence Erlbaum Publishers.

Wickens, C. D. (2004a). Imperfect automation and CDTI Alerting. Paper presented at the Aviation Space and Medical Association (ASMA) annual meeting.

Wickens, C. D. (2005a). Multiple resource time sharing models. In N. Stanton, A. Hedge, K. Brookhuis, and E. Salas (Eds.), *Handbook of Human Factors and Ergonomics Methods* (pp. 40.1–40.6). Boca Raton, FL: CRC Press.

Wickens, C. D. (2005b). Attentional tunneling and task management. AHFD-05-23/ NASA-05-10, University of Illinois, Aviation Human Factors Division.

Wickens, C. D. (2007). Attention to attention and its applications: A concluding view. In Kramer, Wiegmann, and Kirlik, pp. 239–250.

Wickens, C. D., Alexander, A. L., Ambinder, M. S., and Martens, M. (2004). The role of highlighting in visual search through maps. *Spatial Vision* 37: 373–388.

Wickens, C. D. and Andre, A. D. (1990). Proximity compatibility and information display: Effects of color, space, and objectness of information integration. *Human Factors* 32: 61–77.

Wickens, C. D. and Baker, P. (1995). Cognitive issues in virtual reality. In W. Barfield and T. Furness (Eds.), *Virtual environments and advanced interface design* (pp. 515–541). New York: Oxford University Press.

Wickens, C. D. and Benel, D. (1982). The development of time-sharing skills. In J. A. S. Kelso (Ed.), *Motor development* (pp. 253–272). New York: John Wiley and Sons.

Wickens, C. D. and Carswell, C. M. (1995). The proximity compatibility principle: Its psychological foundation and relevance to display design. *Human Factors* 37(3): 473–494.

Wickens, C.D. and Colcombe, A. (2007a). Effects of CDTI alerting system properties on pilot multi-task performance. In *Proceedings 2007 International Symposium on Aviation Psychology*. Dayton: Wright State University Press (CDRom).

Wickens, C. D. and Colcombe, A. (2007b). *Performance consequences of imperfect alerting automation associated with a cockpit display of traffic information, Human Factors,* 49.

Wickens, C. D. and Dixon, S.R. (2007). The benefits of imperfect automation: A synthesis of the literature. *Theoretical Issues in Ergonomics Sciences,* 8(3): 201–212.

Wickens, C.D. and Hollands, J. (2000). *Engineering psychology and human performance* (3rd ed.). Upper Saddle River, NJ: Prentice Hall.

Wickens, C. D., Dixon, S. R., and Ambinder, M. S. (2006). Workload and automation reliability in unmanned air vehicles. In N. J. Cooke, H. Pringle, H. Pedersen, and O. Connor (Eds.), *Advances in human performance and cognitive engineering research,* vol. 7, *Human factors of remotely operated vehicles* (pp. 209–222). Amsterdam, NL: Elsevier Ltd.

Wickens, C. D., Dixon, S., Goh, J., and Hammer, B. (2005). Pilot dependence on imperfect diagnostic automation in simulated UAV flights: An attentional visual scanning analysis. In R. Jensen (Ed.). *Proceedings of the 13th International Symposium on Aviation Psychology*, Wright-Patterson AFB, Dayton OH.

Wickens, C. D., Dixon, S., and Seppelt, B. (2002). *In-vehicle displays and control task interferences: The effects of display location and modality.* Savoy: University of Illinois, Aviation Research Lab.

Wickens, C. D., Dixon, S. R., and Seppelt, B. (2005). Auditory preemption versus multiple resources: Who wins in interruption management? In *Proceedings of the 49th Annual Meeting of the Human Factors and Ergonomics Society* (CDRom). Santa Monica, CA: Human Factors.

Wickens, C. D., Goh, J., Helleberg, J., Horrey, W., and Talleur, D. A. (2003). Attentional models of multi-task pilot performance using advanced display technology. *Human Factors* 45(3): 360–380.

Wickens, C. D., Helleberg, J., and Xu, X. (2002). Pilot maneuver choice and workload in free flight. *Human Factors* 44: 171–188.

Wickens, C. D. and Hollands, J. G. (2000). *Engineering psychology and human performance,* 3d ed. Upper Saddle River, NJ: Prentice Hall.

Wickens, C. D. and Horrey, W. (in press). Models of attention, distraction, and highway hazard avoidance. In M. Reagan, J. D. Lee, and K. Young (Eds.), *The Distracted Driver.*

Wickens, C. D. and Kessel, C. (1980). The processing resource demands of failure detection in dynamic systems. *Journal of Experimental Psychology: Human Perception and Performance* 6: 564–577.

Wickens, C. D., Kramer, A. F., Vanesse, L., and Donchin, E. (1983). The performance of concurrent tasks: A psychophysiological analysis of the reciprocity of information processing resources. *Science* 221: 1080–1082.

Wickens, C. D., Lee, J. D., Liu, Y., and Gordon Becker, S. E. (2004). *An introduction to human factors engineering,* 2d ed. Upper Saddle River, NJ: Pearson Education.

Wickens, C. D. and Liu, Y. (1988). Codes and modalities in multiple resources: A success and a qualification. *Human Factors* 30: 599–616.

Wickens, C. D., Mavor, A. S., Parasuraman, R., and McGee, J. P. (1998). *The future of air traffic control: Human operators and automation.* Washington, DC: National Academy Press.

Wickens, C. D., McCarley, J. S., Alexander, A., Thomas, L., Ambinder, M., and Zheng, S. (2007). Attention-situation awareness (A-SA) model of pilot error. In D. Foyle and B. Hooey (Eds.), *Pilot performance models* (pp. 213–240). Mahwah, NJ: Lawrence Erlbaum.

Wickens, C. D., Merwin, D. H., and Lin, E. (1994). Implications of graphics enhancements for the visualization of scientific data: Dimensional integrality, stereopsis, motion, and mesh. *Human Factors* 36(1): 44–61.

Wickens, C. D., Mountford, S. J., and Schreiner, W. S. (1981). Multiple resources, task-hemispheric integrity, and individual differences in time-sharing efficiency. *Human Factors* 22: 211–229.

Wickens, C. D. and Rose, P. (2001). *Human Factors Handbook for Displays.* Walnut Creek, CA: Rockwell Sciences.

Wickens, C. D. and Sandry, D. (1982). Task-hemispheric integrity in dual task performance. *Acta Psychologica* 52: 227–247.

Wickens, C. D., Sandry, D., and Vidulich, M. (1983). Compatibility and resource competition between modalities of input, output, and central processing. *Human Factors* 25: 227–248.

Wickens, C. D., Sebok, A., Bagnall, T., and Kamienski, J. (2007). Modeling situation awareness supported by advanced flight deck displays. *Proceedings 2007 Conference of the Human Factors and Ergonomics Society,* Santa Monica, CA, Human Factors (CDRom).

Wickens, C. D. and Seidler, K. (1997). Information access in a dual task context. *Journal of Experimental Psychology: Applied* 3: 196–215.

Wickens, C. D., Thomas, L. C., and Young, R. (2000). Frames of reference for display of battlefield terrain and enemy information: Task-display dependencies and viewpoint interaction use. *Human Factors* 42(4): 660–675.

Wickens, C. D., Ververs, P., and Fadden, S. (2004). Head-up display design. In D. Harris (Ed.), *Human factors for civil flight deck design* (pp. 103–140). City, UK: Ashgate.

Wickens, C. D., Vidulich, M., and Sandry-Garza, D. (1984). Principles of S-C-R compatibility with spatial and verbal tasks: The role of display-control location and voice-interactive display-control interfacing. *Human Factors* 26: 533–543.

Wickens, C. D., Vincow, M., Schopper, R., and Lincoln, J. (1997). *Human Performance Models for Display Design*. Wright Patterson AFB: Crew Stations Ergonomics Information Analysis Center SOAR.

Wickens, C. D., Vincow, M., and Yeh, M. (2005). Design applications of visual spatial thinking: The importance of frame of reference. In A. Miyaki and P. Shah (Eds.), *Handbook of visual spatial thinking* (pp. 383–425). Oxford: Oxford University Press.

Wickens, C. D. and Weingartner, A. (1985). Process control monitoring: The effects of spatial and verbal ability and current task demand. In R. Eberts and C. Eberts (Eds.), *Trends in ergonomics and human factors*. Amsterdam: North Holland Publishing Co.

Wiegmann, D., McCarley, J. S., Kramer, A. F., and Wickens, C. D. (2006). Age and automation interact to influence performance of a simulated luggage screening task. *Aviation Space and Environmental Medicine* 77(8): 825–831.

Wierwille, W. and Casali, J. J (1983). A validated rating scale for global mental workload measurement applications. In *Proceedings 27th Annual Meeting of the Human Factors Society* (pp. 25–32). Santa Monica, CA: Human Factors.

Williams, L. J. (1982). Cognitive load and the functional field of view. *Human Factors,* 24, 683–692.

Wightman, D. C. and Lintern, G. (1985). Part task training for tracking and manual control. *Human Factors* 27: 267–283.

Wilkinson, L. (1999). Statistical methods in psychology journals. *American Psychologist* 54: 594–603.

Wilkinson, R. T. (1964). Artificial "signals" as an aid to an inspection task. *Ergonomics* 7: 63–72.

Williams, A. M. and Davids, K. (1998). Visual search strategy, selective attention, and expertise in soccer. *Research Quarterly for Exercise and Sport* 69: 111–128.

Williams, H., Wickens, C. D., and Hutchinson, S. (1996). A comparison of methods for promoting geographic knowledge in simulated aircraft navigation. *Human Factors* 38(1): 50–64.

Williams, H. P., Hutchinson, S., and Wickens, C. D. (1996). A comparison of methods for promoting geographic knowledge in simulated aircraft navigation. *Human Factors* 38(1): 50–64.

Williams, L. J. (1982). Cognitive load and the functional field of view. *Human Factors* 24: 683–692.

Williges, R. and Wierwille, W. (1979). Behavioral measures of aircrew mental workload. *Human Factors* 21: 549–574.

Wingfield, A., Tun, P. A., and McCoy, S. L. (2005). Hearing loss in older adulthood: What it is and how it interacts with cognitive performance. *Current Directions in Psychological Science* 14: 144–148.

Wogalter, M. S. and Laughery, K. R. (2006). Warnings and hazard communications. In Salvendy, pp. 889–911.

Woldorff, M. G. and Hillyard, S. A. (1991). Modulation of early auditory processing during selective listening to rapidly present tones. *Electroencephelography and clinical neurophysiology* 79: 170–191.

Wolf, L. D., Potter, P., Sedge, J., Bosserman, S., Grayson, D., and Evanoff, B. (2006). Describing nurses' work: Combining quantitative and qualitative analysis. *Human Factors* 48: 5–14.

Wolfe, J. M. (1994). Guided search 2.0: A revised model of visual search. *Psychonomic Bulletin and Review* 1: 202–238.

Wolfe, J. M. (1998). Visual search. In H. Pashler (Ed.), *Attention* (pp. 13–73). East Sussex: Psychology Press.

Wolfe, J. M. (2001). Asymmetries in visual search: An introduction. *Perception and Psychophysics* 63: 381–389.

Wolfe, J. M., Cave, K. R., and Franzel, S. L. (1989). Guided search: An alternative to feature integration model for visual search. *Journal of Experimental Psychology: Human Perception and Performance* 15: 419–433.

Wolfe, J. M. and Horowitz, T. S. (2004). What attributes guide the deployment of visual attention and how do they do it? *Nature Neuroscience* 5: 1–7.

Wolfe, J. M., Horowitz, T. S., and Kenner, N. M. (2005). Rare items often missed in visual searches. *Nature* 435: 439–440.

Wood, N. L. and Cowan, N. (1995). The cocktail party phenomenon revisited: Attention and memory in the classic selective listening procedure of Cherry (1953). *Journal of Experimental Psychology: General* 124: 243–262.

Woodman, G. F. and Luck, S. J. (1999). Electrophysiological measurement of rapid shifts of attention during visual search. *Nature, 400,* 867–869.

Woodman, G. F. and Luck, S. J. (2003). Serial deployment of attention during visual search. *Journal of Experimental Psychology: Human Perception and Performance* 29: 121–138.

Woods, D. D. (1984). Visual momentum: A concept to improve the coupling of person and computer. *International Journal of Man-Machine Studies* 21: 229–244.

Woods, D. D. (1995). The alarm problem and directed attention in dynamic fault management. *Ergonomics* 38(11): 2371–2393.

Woods, D. D, Wise, J., and Hanes, L. (1981). An evaluation of nuclear power plant safety parameter display systems. *Proceedings of the 25th Annual Meeting of the Human Factors Society,* Santa Monica, CA: Human Factors.

Woollacott, M. and Shumway-Cook, A. (2002). Attention and the control of posture and gait: A review of an emerging area of research. *Gait and Posture* 16: 1–14.

Wright, D. B. and Davies, G. M. (2007). Eye witness testimony. In Durso et al., pp. 763–788.

Xu, X. and Rantanen, E. (2007) Effect of air traffic geometry on pilot's conflict detection with a cockpit display of traffic information. *Human Factors, 46,* 358–375.

Xu, X., Wickens, C. D., and Rantanen, E. M. (2007). Effects of conflict alerting system reliability and task difficulty on pilots' conflict detection with cockpit display of traffic information. *Ergonomics* 50, 112–130.

Yantis, S. (1993). Stimulus-driven attentional capture. *Current Directions in Psychological Sciences* 2: 156–161.

Yantis, S. and Egeth, H. E. (1999). On the distinction between visual salience and stimulus-driven attentional capture. *Journal of Experimental Psychology: Human Perception and Performance* 25: 661–676.

Yantis, S. and Hillstrom, A. P. (1994). Stimulus-driven attentional capture: Evidence from equiluminant visual objects. *Journal of Experimental Psychology: Human Perception and Performance, 20,* 95–107.

Yantis, S. and Johnston, J. C. (1990). On the locus of visual selection: Evidence from focused attention tasks. *Journal of Experimental Psychology: Human Perception and Performance* 16: 135–149.

Yantis, S. and Jonides, J. (1984). Abrupt visual onsets and selective attention: Evidence from visual search. *Journal of Experimental Psychology: Human Perception and Performance, 10,* 601–621.

Yantis, S. and Jonides, J. (1990). Abrupt visual onsets and selective attention: Voluntary versus automatic allocation. *Journal of Experimental Psychology: Human Perception and Performance, 16,* 121–134.

Yeh, M. and Wickens, C. D. (2001a). Display signaling in augmented reality: The effects of cue reliability and image realism on attention allocation and trust calibration. *Human Factors* 43(3): 355–365.

Yeh, M. and Wickens, C. D. (2001b). Attentional filtering in the design of electronic map displays: A comparison of color-coding, intensity coding, and decluttering techniques. *Human Factors* 43(4): 543–562.

Yeh, M., Merlo, J., Wickens, C. D., and Brandenburg, D. L. (2003). Head up versus head down: The costs of imprecision, unreliability, and visual clutter on cue effectiveness for display signaling. *Human Factors* 45(3): 390–407.

Yeh, M., Wickens, C. D., and Seagull, F. J. (1999). Target cueing in visual search: The effects of conformality and display location on the allocation of visual attention. *Human Factors* 41(4): 524–542.

Yeh, Y. Y. and Wickens, C. D. (1988). Dissociation of performance and subjective measures of workload. *Human Factors* 30: 111–120.

Young, M. S. and Stanton, N. A. (2002). Malleable attentional resources theory: A new explanation for the effects of mental underload on performance. *Human Factors* 44: 365–375.

Subject Index

227

Author Index